Framing a Strategic Approach for

Reserve Component Joint Officer Management

W0017607

Harry J. Thie, Margaret C. Harrell, Sheila Nataraj Kirby,
Al Crego, Roland J. Yardley, Sonia Nagda

Prepared for the
Office of the Secretary of Defense

NATIONAL DEFENSE RESEARCH INSTITUTE

The research described in this report was prepared for the Office of the Secretary of Defense (OSD). The research was conducted in the RAND National Defense Research Institute, a federally funded research and development center sponsored by the OSD, the Joint Staff, the Unified Combatant Commands, the Department of the Navy, the Marine Corps, the defense agencies, and the defense Intelligence Community under Contract DASW01-01-C-0004.

Library of Congress Cataloging-in-Publication Data

Thie, Harry.
    Framing a strategic approach for reserve component joint officer management /
Harry J. Thie, Margaret C. Harrell, [et al.].
      p. cm.
    Includes bibliographical references.
    ISBN-13: 978-0-8330-3973-6 (pbk. : alk. paper)
      1. United States—Armed Forces—Officers—Management. 2 United States—
Armed Forces—Reserves—Management. 3. Unified operations (Military science)
I. Harrell, Margaret C. II. Title.

UB413.T47 2007
355.3'32—dc22

                                       2006033293

The RAND Corporation is a nonprofit research organization providing objective analysis and effective solutions that address the challenges facing the public and private sectors around the world. RAND's publications do not necessarily reflect the opinions of its research clients and sponsors.

RAND® is a registered trademark.

Cover design by Stephen Bloodsworth

Published 2006 by the RAND Corporation
1776 Main Street, P.O. Box 2138, Santa Monica, CA 90407-2138
1200 South Hayes Street, Arlington, VA 22202-5050
4570 Fifth Avenue, Suite 600, Pittsburgh, PA 15213-2665
RAND URL: http://www.rand.org/
To order RAND documents or to obtain additional information, contact
Distribution Services: Telephone: (310) 451-7002;
Fax: (310) 451-6915; Email: order@rand.org

Preface

The 2005 National Defense Authorization Act calls on the Secretary of Defense to "develop a strategic plan for joint officer management and joint professional military education that links joint officer development to the accomplishment of the overall missions and goals of the Department of Defense . . . for the purpose of ensuring that sufficient numbers of officers fully qualified in occupational specialties involving combat operations are available as necessary to meet the needs of the Department for qualified officers who are operationally effective in the joint environment."

An earlier RAND Corporation project framed a strategic approach for joint officer management in the active component. The current project, which builds on the earlier effort, seeks to (a) operationalize this strategic approach for joint officer management in the active component through extensive data analysis and complex modeling, and (b) develop and frame a strategic approach to joint officer management in the reserve component.

This report documents the work done for the latter task. It provides an overview of the current approach to joint officer management for the reserve component and the constraints unique to the reserve component, compares and contrasts the officer development processes of active component and reserve component officers, explores how demand for joint reserve component officers might be determined and the various ways that reserve component officers receive joint experience, and determines alternatives to the current system of determining which officers are validly joint. The overall aim of the project task and of the report is to develop a conceptual plan for a strategic approach to reserve officer development in joint matters, grounded in lessons learned about effective human resource strategies, officer management, and joint officer matters.

This research was sponsored by the Undersecretary of Defense for Personnel and Readiness. It was conducted within the Forces and Resources Policy Center of RAND's National Defense Research Institute, a federally funded research and development center sponsored by the Office of the Secretary of Defense, the Joint Staff, the Unified Combatant Commands, the Department of the Navy, the Marine Corps, the defense agencies, and the defense Intelligence Community. The principal investigators were Harry Thie and Margaret Harrell. Comments are welcome and may be addressed to Harry Thie at harry_thie@rand.org or to Margaret Harrell at margaret_harrell@rand.org.

For more information on RAND's Forces and Resources Policy Center, contact the director, James Hosek. He can be reached by email at james_hosek@rand.org; by phone at 310-393-0411, extension 7183; or by mail at the RAND Corporation, 1776 Main Street, Santa Monica, CA 90407-2138. More information about RAND is available at www.rand.org.

Contents

Figures

Tables

Acronyms

AC	Active Component
AFMPC	Air Force Manpower and Personnel Center
AFR	Air Force Reserve
AGR	Active Guard/Reserve
AJPME	Advanced Joint Professional Military Education
ANG	Air National Guard
ARNG	Army National Guard
ASD(RA)	Assistant Secretary of Defense for Reserve Affairs
CENTCOM	United States Central Command
CJCS	Chairman of the Joint Chiefs of Staff
CONUS	Continental United States
CSA	Combat Support Agencies
DLA	Defense Logistics Agency
DoD	Department of Defense
DoDI	Department of Defense Instruction
DOPMA	Defense Officer Personnel Management Act
DUSD(PI)	Deputy Under Secretary of Defense for Program Integration
FJQ	fully joint qualified
GFO	General and Flag Officer
GNA	Goldwater-Nichols Act (of 1986)
GWOT	Global War on Terror

HR	Human Resources
ICAF	Industrial College of the Armed Forces
IMA	Individual Mobilization Augmentee
JCWS	Joint and Combined Warfighting School
JDA	Joint Duty Assignment
JDAL	Joint Duty Assignment List
JDA-R	Joint Duty Assignment-Reserve
JFCOM	Joint Forces Command
JFSC	Joint Forces Staff College
JIT	Joint Individual Training
JOD	Joint Officer Development
JOM	Joint Officer Management
JPME	Joint Professional Military Education
JPME I	First Phase of Joint Professional Military Education
JPME II	Second Phase of Joint Professional Military Education
JQO	Joint Qualified Officer
JRD	Joint Reserve Directorate
JRU	Joint Reserve Unit
JS	Joint Staff
JSO	Joint Specialty Officer
JTF	Joint Task Force
JTF HQS	Joint Task Force Headquarters
KSA	Knowledge, Skills, and Abilities
KSABQ	Knowledge, Skills, and Abilities-Based Questionnaire
LMI	Logistics Management Institute
MILPDS	Military Personnel Data System
MOS	military occupational specialty
MQ	Minimum Qualification
NDAA	National Defense Authorization Act

NMS	National Military Strategy
NORTHCOM	United States Northern Command
NWC	National War College
OPMEP	Officer Professional Military Education Policy
OSD	Office of the Secretary of Defense
PAF	Project Air Force (RAND)
PDS	Personnel Data System
PME	Professional Military Education
RASL	Reserve Active Status List
RC	Reserve Component
RCCPDS	Reserve Component Common Personnel Data System
RFPB	Reserve Forces Policy Board
SJFHQ	Standing Joint Force Headquarters
SME	Subject Matter Experts
T&E	Training and Experience
TAD/TDY	Temporary Assigned Duty/Temporary Duty
TBQ	Task-Based Questionnaire
UIC	Unit Identification
UOR	Uniform Officer Record
USAFR	United States Air Force Reserve
USD(P&R)	Under Secretary of Defense for Personnel and Readiness

Summary

The Department of Defense (DoD) management processes for active component joint duty assignments, education, and training were built around the solid foundation provided by the Goldwater-Nichols Act (GNA) of 1986. However, it is being increasingly recognized that the current approach to joint matters needs to evolve from its current static format to a more dynamic approach that broadens the definitions of "joint matters" and "joint qualifications" and allows for multiple paths to growing the number of joint officers. An important extension of the current strategic plan is a more explicit and strategic consideration of reserve component joint officer management. The need for a systematic examination of how reserve component joint officers are trained and developed is becoming increasingly urgent, given the dramatic increase in the use of the reserve forces.

Building on work done earlier for the active component with respect to joint officer management, this research focuses on framing a strategic approach to reserve joint officer management that (a) addresses the requirements for and the supply of joint officers for the reserve component and (b) accounts for the unique constraints of and challenges to reserve joint officer management. A strategic approach for reserve component joint officer management must deliberately determine which jobs require joint experience or which provide it. In particular, given the current strategic intent of the DoD with respect to jointness ("push it to its lowest appropriate level"), the need for joint experience should be measurable in a much larger number of billets, in particular in billets internal to the service. Moreover, valid joint experience might now be provided by service in billets internal to the services, particularly those associated with Joint Task Forces (JTFs), with service component commands, and with joint planning and operations.

Producing Joint Officers

For active duty officers, a joint duty assignment (JDA) is defined as one in which the officer gains significant experience in joint matters. A list of such assignments, called the Joint Duty Assignment List (JDAL), is maintained. Moreover, if an officer gains education and experience of particular types for specified durations, he becomes a Joint Specialty Officer (JSO), or someone who is trained in and oriented toward joint matters. If an officer has completed the second phase of joint professional military education (JPME II) but not a JDA, he is considered a JSO nomination. The DoD instruction on reserve component joint officer management issued in 2002 introduces two additional terms to the lexicon, *Fully Joint Qualified* and *Joint Officer*. The first requires Advanced Joint Professional Military Education (AJPME) (the reserve component [RC] equivalent of JPME II) and sufficient time in a qualifying billet. This is not unlike a JSO. The second term defines an officer who has achieved Chairman of the Joint Chiefs of Staff (CJCS) joint learning objectives[1] and has served or shall serve in a joint duty assignment-reserve (JDA-R) billet. For one who has served in a JDA-R billet, the difference between being a Joint Officer and Fully Joint Qualified appears to be in achieving CJCS learning objectives through a process other than AJPME.

---

[1] The instruction does not define how this is measured. It could be through school attendance or self-study or other means.

Determining Demand for Reserve Active-Status List Officers with Prior Joint Education and/or Experience

The current level of utilization of the RC and new missions, organizations, and structures stood up within the RC point to an increased recognition that the work required of reservists is becoming increasingly joint.[2] Like the JDAL for active duty list officers, the JDA-R is focused on the supply of officers with joint experience. Both lists designate positions that qualify officers with joint experience; neither shows where such experience might be needed. Although some services and components may have begun to make these determinations, we found no databases that routinely collected it.

Demand for prior jointness—joint education and experience—is likely to be very different across the different reserve categories. For example, the typical reserve unit staffed by drilling reservists is unlikely to have a demand for officers educated and trained in joint matters, with perhaps a few exceptions. More than 70 percent of O-4 and above selected reservists are in this category. The exceptions could be unit reservists who may volunteer for active duty for special work (for example, in a JTF) or Individual Mobilization Augmentees (IMAs) who are generally assigned to active component organizations including positions in joint organizations. This is also true of Active Guard/Reserve (AGR) officers who are serving in external organization billets, for example, at the Office of the Secretary of Defense (OSD) or on the Joint Staff or in combatant commands. So we would expect that much of the demand for IMAs and AGRs with joint experience or education would be a derived demand from active component organizations, external to the military service, with little demand within service reserve component units.

Demand for prior jointness is also likely to be very different across the different reserve components. For example, Navy reserve officers

---

[2] This is a result of several factors to include more military work being defined as joint, more service work being done in a joint operating environment, and reservists participating in such work and in greater numbers. Moreover, homeland security and homeland defense tasks (HLS/HLD) are increasing, especially with respect to interagency and intergovernmental needs.

have few opportunities to serve as IMAs in Navy organizations or in unit augmentations to Navy active component units. And thus the Navy Reserve is less likely to require or provide its officers with joint experience. However, Guard units have both federal and state missions and may be called up to coordinate homeland defense or disaster relief with a number of state and federal agencies. Thus state missions are more likely to provide interagency and intergovernmental joint experience and to benefit from such prior joint experience.

Indeed, demand for prior jointness is also likely to be very different across the different kinds of jointness. For active duty list officers, the greatest demand is for multiservice qualified officers, then multinational, and last interagency qualified officers. For the reserve components, especially the Army and the Air National Guard, interagency qualifications are likely to be in far greater demand than multinational qualifications. Moreover, intergovernmental experience at the state and local level also plays a role in demand for the National Guard that is not typically seen for active duty list officers.

Determining Potential Supply of Reserve Joint Officers

We recognize that management of reserve active-status list officers is more of a "pull" system than a "push" system. There are not central assignment processes as there are for active duty list officers. As a result, individual officers must be heavily involved in decisions about their assignments and education. The concept of a formal opt-in system of reserve joint officer management is to be explored. Particularly, this might be useful for those who aspire to general or flag officer. Moreover, the assigning and educating processes, to the extent they exist, vary for the different reserve categories: unit-based officers, individual mobilization augmentees, and active guard and reserve.

The debate about how to obtain joint credit for both active duty list and reserve active-status list officers centers on which circumstances, whether formal assignment or ad hoc responsibilities, provide a valid joint experience for officers. The active component system is currently constrained by the GNA (as contained in Title 10 U.S. Code) as far as the types of assignments (and the tenure required in those assignments) before an officer receives joint credit. Reserve component joint officer management (JOM) is not so constrained. Therefore, in our analysis, we explored a number of different systems that could be used for evaluating either processes or outputs/outcomes as a basis for qualifying officers—methods that provide the building blocks for a more flexible, accommodating, and valid joint officer management system.

If a billet- and time-based structure serves as the primary element for identifying joint experience, then the JDA-R could contain those assignments that are judged to consistently provide each officer who performs that job with a valid joint experience. Such a time- and billet-based system is administratively simple, albeit relatively inflexible. The required tenure might be consistent for all jobs, or it might vary by location or by characteristics of the job. However, there would be less pressure for the JDA-R to include all billets that might possibly, under some conditions, provide a joint experience, as the billet-based system could be complemented by an individual evaluation method that acknowledges the joint experience gained by officers in other, non–JDA-R assignments.

An individual-based system could provide joint credit to those officers who documented their proficiency in identified joint areas. This system might consider the valid joint experiences of officers who were serving in civilian positions[3] or positions in organizations external to the service that were not on the JDA-R, or who served in JDA-R assignments for less than the identified required tenure. Such a system might also acknowledge the valid joint experiences of officers who were serving in service-specific organizations if such billets were not on the JDA-R. The evaluation criteria that assess the depth or breadth of experience required might vary for those officers serving in internal service positions, or the level of authority required to approve the joint experience of officers in internal service assignments might vary from those in external organizations; but this system could flexibly acknowledge officers who received a valid joint experience in assignments that were otherwise not typically or consistently joint. To the extent that officers

---

[3] DoD surveys reservists to determine their civilian skills.

in certain assignments were consistently applying for, and receiving, joint credit, such assignments should be considered for addition to the JDA-R.[4]

Regardless of the combination of accreditation structures used to identify officers who have received a valid joint experience, joint education, joint training, or even joint acculturation, there should be recognition in the system of different levels of joint proficiency. For example, while it is important to acknowledge those officers who are fully joint qualified, it is also important to recognize those officers who have received sufficient joint education to begin a joint position with some proficiency. Such proficiency levels could also acknowledge the relatively different levels of experience gained by reserve active-status list officers. These levels might be joint officer or fully joint qualified officers. Additionally, there may be value in separately identifying those officers who are proficient and experienced in multinational issues, multiservice issues, interagency issues, and intergovernmental issues. Doing so, however, should be decided after an assessment of the number of officers needed with these particular backgrounds.

Estimating Supply of Reserve Joint Officers

Approximately 225 reserve active-status list officers have completed AJPME through January 2006, about 250 will complete AJPME in 2006, and another 400 will complete AJPME in 2007. Thereafter, the Joint Forces Staff College (JFSC) plans to provide AJPME for as many as 500 reserve active-status list officers on an annual basis. We cannot, however, estimate with current data the number of reserve officers who have obtained a joint experience or joint education and who remain reserve officers. It is likely that there is currently an insufficient amount of reserve active-status list officers with joint experience and education, even if they are tracked and assigned appropriately to maximize the utilization of reservists who might be considered joint officers, or fully

---

[4] The process and outcome of providing joint credit to officers in an internal service organization should be carefully monitored and assessed, perhaps in a pilot study to determine what kind of officers were receiving joint credit and for what kind of job experiences. Moreover, the discussion, while focused on JDA-R and non–JDA-R positions is applicable to JDAL and non-JDAL positions for active duty list officers.

joint qualified, were such assessment practices in place. The relatively large demand for reservists[5] suggests that there is a need to assess the joint qualification level of reserve active-status list officers to ensure that the relatively scarce resource of reservists with joint qualifications are managed for the best utilization of those qualifications. An important step is to verify that there is an actual or incipient shortage of joint-qualified reserve officers.

Conclusions and Recommendations

A framework of law and policy is in place for joint officer management. For active duty list officers, much of that framework is in law; for reserve active-status list officers, most of the framework is in policy. A strategic approach to joint officer management for reserve active-status list officers must assess the need for officers with prior joint knowledge, joint experience, and acculturation before assignment to certain positions. Those positions are not yet identified, nor are the needs for officers with multinational, multiservice, interagency, or intergovernmental knowledge and experience. Given identified needs, a strategic approach looks at the current inventory of available officers and projects the future availability given qualification, assignment durations, promotion, and retention rates. A current documented inventory of available officers does not exist. There are many officers who, because of current deployments and employment, have gained joint training and experience, but these qualifications are not visible within any data system.

---

[5] The RC's joint demand is likely to vary widely over time. There is currently a large demand for the reserve component, some of it joint demand, and it may be a long-term demand. The active component has a (more or less) constant force structure with constant joint demands, but the RC is a "surge" force in transition (rebalancing more than the active force) that may continue to be more of an operational force. Even in its more "operational" form, the percent serving on active duty may vary from 25 percent (now), to 10 percent (pre-9/11), to something between in the future. Analysis within a demand and supply framework should allow for demand and supply (qualification) to vary depending on joint use.

Recommendations

We offer several recommendations to help implement a strategic approach to reserve component joint officer management.

- First, data are needed about requirements and should be collected. The services should be directed to incorporate data about the needs for prior joint education, or prior joint experience, and thus for fully joint qualified officers and joint officers, into their manpower databases. At the same time, the Joint Staff (JS) should implement and maintain a stand-alone database for at least the JS, OSD, combatant commands, and other external organizations until such time that the manpower systems are changed. A one-time data collection[6] should be done to populate this database initially with updates as needed. In particular, the latter two sources of demand could vary most for the reserve component over time as they surge to meet changing geostrategic situations.
- Second, a supply-oriented database such as the JDA-R documenting the positions that provide joint experience should also be populated. Procedures for doing this for the JDA-R are laid out in the DoD Instruction and should be implemented.
- Third, OSD, the JS, and the services should specify policies and procedures for capturing information about qualifying knowledge and the experience of officers beyond that which will be captured by the billet- and time-based JDA-R system. We recommend that a points system or an accomplishment record approach be used, as discussed in more detail in Chapter Five.
- Fourth, the future supply of qualified reserve active-status list officers should be projected using modeling of JFSC AJPME seats, JDA-R positions, assignment duration, qualification by other means, and likely promotion and retention rates.
- Both current and projected inventory needs to be compared with the demand to determine where shortages and overages exist as a basis for formulating appropriate policy alternatives. This analysis

---

[6] This process is outlined in Appendices B and C of *Framing a Strategic Approach for Joint Officer Management* (MG-306-OSD) and a similar process could be followed.

would determine the extent to which the need for officers with joint education and experience can be satisfied by the number of qualifying billets, other qualifying experiences, and educational seats combined with the use, promotion, and other management practices for officers of different reserve categories and occupational communities.[7] The strategic plan should lastly determine the policies and practices to align the amount of jointness available with the demand for jointness.

- Finally, the implemented strategic approach, which recognizes both the need for jointness among reserve active-status list officers and the complementary means to acknowledge and accredit joint qualifications for reserve officers, should be evaluated and considered for its application to active duty list officers.

---

[7] Because of continuing change in the roles and missions of the reserve components, these assessments will need to be made periodically, if not continually, to avoid long-term management problems.

Acknowledgments

This research benefited from the contributions of many individuals—active duty list and reserve active-status list—on service staffs, in combatant commands, in service component commands, in the Office of the Secretary of Defense and the Joint Staff, and in senior and intermediate staff colleges who spoke candidly with us regarding reserve component joint officer personnel management and service policy and attitudes regarding reserve component joint issues.

At risk of omitting someone, we are particularly grateful for the support and interaction of Rich Krimmer, Lieutenant Colonel Charles Armentrout, Lieutenant Colonel Susan Hogg, Brad Loo, and Lieutenant Colonel Ken Walters from our sponsoring offices. Navy Captain Steve Elrod, Lieutenant Colonel Judi Davenport, Kenneth Pisel, Colonel Gary Harper, Colonel Alison Jameson, and Major Christine Klink were particularly insightful on these issues. RAND colleagues Holly Potter, Lieutenant Colonel Eric Haaland, Major Brian Maue, Michael Tseng, and Peter Schirmer provided significant help.

We are indebted to our reviewers, Hans Pung and Dallas Owens, for their suggestions that improved the report. We also appreciate the administrative support provided by Vicki Wunderle.

CHAPTER ONE

Introduction

Background

The Goldwater-Nichols Act (GNA) of 1986 forged a cultural revolution in the U.S. Armed Forces by improving the way the Department of Defense (DoD) prepares for and executes its mission. Title IV of the GNA addresses joint officer personnel policies and provides specific personnel management requirements for the identification, education, training, promotion, and assignment of officers into joint duties. The DoD management processes for joint duty assignments, education, and training were built around the solid foundation provided by GNA. In the past 15 years, successes in Iraq (Desert Shield/Storm), Bosnia, and Afghanistan (among others), and more recently in Operation Iraqi Freedom, testify to the effectiveness of the joint military force and its war-fighting potential. However, counterinsurgency and irregular warfare are becoming more central. It is being increasingly recognized that the current approach to joint matters needs to evolve from its current static format to a more dynamic approach that broadens the definitions of "joint matters" and "joint qualifications" and allows for multiple paths to growing the number of joint officers.[1]

---

[1] General Accounting Office, *Joint Officer Development Has Improved, but a Strategic Approach Is Needed,* Washington, D.C., 2002; Booz Allen Hamilton, *Independent Study of Joint Officer Management and Joint Professional Military Education,* McLean, VA, 2003.

For example, a 2002 study by the General Accounting Office (renamed the Government Accountability Office in July 2004) conducted an assessment of DoD actions to implement provisions in law that address the development of officers in joint matters and concluded that "a significant impediment affecting DoD's ability to fully realize the cultural change that was envisioned by the act is the fact that DoD has not taken a strategic approach to develop officers in joint matters."[2] In March 2003, an independent study[3] authorized under the 2002 National Defense Authorization Act (NDAA) for 2002 indicated that joint officer management and the professional military education afforded to joint officers require updating in practice, policy, and law to better meet the demands of a new era more effectively, and that such changes should be undertaken as part of an overall strategic approach to developing the officer corps for joint warfare. DoD's most recent strategic plan for joint officer management and joint officer development states eloquently:

> [T]he military is "joint" by operational necessity . . . Joint Task Forces (JTFs) now define the way we array our armed forces for war and operations other than war. Jointness is no longer simply the integration and/or interoperability of two or more military services; it involves the synergistic employment of multi-component forces from multiple services, agencies, and nations. Non-governmental agencies and commercial enterprises must now be routinely combined with traditional military forces to achieve national objectives. Such a dynamic and varied environment demands flexibility, responsiveness, and adaptability not only from the individual soldiers, sailors, airmen, and marines, but *also from the processes which support them*.[4]

An important extension of the current strategic plan is a more explicit and strategic consideration of reserve component joint officer management. The need for a systematic examination of how reserve

---

[2] General Accounting Office, ibid.

[3] Booz Allen Hamilton, ibid.

[4] U.S. Department of Defense, *Strategic Plan for Joint Officer Management and Joint Officer Development*. April 3, 2006, pp. 2–3.

active-status list officers are trained and developed in joint matters is becoming increasingly urgent, given the dramatic increase in the use in the reserve forces. Since September 11, 2001, almost half a million reservists have participated in the global war on terrorism,[5] well over 4,000 reserve component officers are serving in joint organizations, and there are 11 general and flag officer positions that—at the discretion of the Chairman of the Joint Chiefs of Staff—can be designated as joint duty assignment positions reserved for reserve component officers at a general or flag officer grade.[6] Reservists are often called up with short notice and are expected to perform at the same level of professionalism and readiness in complex joint environments as their active component counterparts, despite having much less time and opportunity to obtain joint professional education or prior joint experience. Despite the requirement in the GNA that the Secretary of Defense establish personnel policies for reserve officers that emphasize education and experience in joint matters,[7] work remains to be done. The purpose of this report is to address these issues.

Purpose of Project

The RAND Corporation has been assisting DoD in framing and operationalizing a strategic approach to joint officer management. The first phase of the project, which focused on active component officers, is documented in Thie, Harrell, et al. (2005).[8] The report outlined a strategic approach for joint officer management and recommendations for operationalizing the strategic plan, including a large survey to collect data on individuals serving in billets that were likely either to require prior joint experience or education or to provide officers with joint experience. This survey—the 2005 Joint Officer Management

---

[5] Cited in U.S. DoD, ibid.

[6] 10 U.S.C., §526 (b)(2)(A).

[7] 10 U.S.C., §666.

[8] Thie, Harrell, et al., ibid.

(JOM) Survey[9]—was fielded in summer 2005 and examined more than 20,000 joint or potentially joint officer billets. The findings are reported in Kirby et al. (2006). Continuing work is examining the extent to which prior jointness is required by billets, and whether sufficient numbers of officers with joint education, training, and experience are likely to be available to satisfy those needs through extensive analysis and modeling of the survey data.

A major task of the current project is to develop a strategic approach for reserve component joint officer management. This latter task was in response to the legislative request for a strategic approach for joint officer management in the reserve component:

> The Secretary of Defense shall develop a strategic plan for joint officer management and joint professional military education that links joint officer development to the accomplishment of the overall missions and goals of the Department of Defense, as set forth in the most recent national military strategy . . . Such plan shall be developed for the purpose of ensuring that sufficient numbers of officers fully qualified in occupational specialties involving combat operations are available as necessary to meet the needs of the Department for qualified officers who are operationally effective in the joint environment. In developing the strategic plan . . . the Secretary shall include joint officer development for officers on the reserve-active status list in the plan.[10]

The strategic approach for reserve joint officer management outlined in this report builds on the strategic approach for joint officer management in the active component, while recognizing that there are considerable differences between officer management and thus joint officer management in the active and reserve components. Reserve joint officer management needs to (1) address the extent to which the requirements for and the supply of joint officers differs for the reserve component and (2) account for the unique constraints of and chal-

---

[9] This is formally titled the JOM Census Survey because it was intended to be a census of all billets. The descriptive write-up of the data is referred to as the Census Survey. It is shortened to Survey in this report.

[10] National Defense Authorization Act for FY 2005.

lenges to reserve component joint officer management.[11] Indeed, the Act itself explicitly recognized the constraints facing the reserve components by stating

> Such policies shall, *to the extent practicable for the reserve components,* be similar to the policies provided by this chapter [for active component officers].[12]

Policies and practices designed for joint officer management in the reserve component—even if different from those designed for active component officers—need not produce "second class joint officers." Indeed, there are possible policies or practices appropriate to the reserve that might offer important lessons for revising active component joint officer management as well.

Although there is general agreement that the laws, policies, and practices pertaining to joint officer management need to be reconsidered and updated to reflect the current environment and operations, there exist very few legacy constraints or rules to reserve component joint officer management. Thus the problems pertaining to reserve joint officer management are not as clearly defined, and the extent to which the perceived problems with active component joint officer management—or recommended repairs to problems—are appropriate to reserve joint officer management is unclear. This is quite simply because the needs of a reserve joint officer management system are less broadly known. The current RC JOM system lacks many constraints; it also lacks positive guidance and implementation.

---

[11] Any of the interviews conducted with joint officers and personnel who manage joint officers suggested the appeal of consistent or even identical (to the extent possible) management guidelines for reserve and active officers. These opinions are generally based on a perception that only identical systems produce equally valued officers. However, these opinions often lack full understanding of the considerable constraints to joint officer management, the quandaries that current policies and laws have produced in the management of active component officers, and the extent to which the same constraints would affect reserve component officer management if applied to the reserve component.

[12] 10 USC, § 666, italics and words in parentheses added.

Current Work on Active Component Joint Officer Management

As mentioned above, this report builds on earlier work that framed a strategic approach to joint officer management in the active component. Thie and Harrell et al. (2005) point out that a strategic approach must understand the need or requirement for critical workforce characteristics and the ability of the management system to provide officers with those characteristics. Moreover, the approach needs to demonstrate (a) a strategy or policy for aligning the availability of officers who have the characteristics with the need for them or (b) a rationale for why more widespread availability of the characteristics rather than the immediate need for them would be desirable. A strategic approach for joint officer management must deliberately determine which jobs, inside or outside the service, require joint experience or provide it. In particular, given the current strategic intent of the Department with respect to jointness ("push it to its lowest appropriate level"), the need for joint experience should be measurable in a much larger number of billets, in particular in billets internal to the service. Moreover, valid joint experience might now be provided by service in billets internal to the service, particularly those associated with Joint Task Forces, with service component commands, and with joint planning and operations. The key components of a strategic approach can be discerned as (a) which jobs require or provide joint experience, (b) how many of each exists, and (c) what is needed to align those two sets of jobs.

Organization of the Report

This report on reserve joint officer management is organized into several chapters. Chapter Two sets the context for the report: It compares and contrasts the officer management and developmental processes in the active and reserve components and then describes the current approach to active and reserve joint officer management (or, in the case of the latter, the lack of one).

Chapter Three outlines a strategic approach for reserve component joint officer management. Because such an approach emphasizes the need to understand demand for and supply of joint officers now and in the future, the next two chapters are devoted to these two topics. Chapter Four considers how demand for jointness among reserve component officers might be determined, first looking at current DoD guidance for determining valid Joint Duty Assignments–Reserve (JDA-R) and then examining various developments in the reserve components that are likely to add to demand for reserve active-status list joint officers. Chapter Five examines the supply of reserve active-status list joint officers and explores how the current approach might be made more flexible to broaden the definition of a valid joint experience and to provide joint professional military education. Both of these chapters provide some limited data from the 2005 JOM Survey on 679 reservists/guardsmen who responded to the survey regarding the characteristics of the billets in which they were serving.

Chapter Six provides conclusions and general insights and recommendations to improve reserve component joint officer management. We conclude that the current reserve data would need to be reinforced with new and more systematic data collection efforts targeted at collecting the kinds of information needed to identify billets that provide joint experience or that need prior joint experience and education.

There are two appendices. Appendix A examines the availability and appropriateness of the data currently available on reserve component officers. Appendix B provides a theoretical framework for the training and experience rating methods referred to in the report and an overview of the different methods currently used.

CHAPTER TWO

Developing and Managing Officers in the Active and Reserve Components

This chapter provides context for the study. The chapter has three sections. The first describes how officers are assessed, developed, and managed generally in the active and reserve components. The second focuses on how active and reserve joint officers are produced. The third section is an assessment of "jointness."

Producing Officers

Reserve Active-Status List and Active Duty List Careers Are Different

There are structural differences between careers of active duty list and reserve active-status list officers that lead to differences in how officers are managed and developed. These structural differences occur in personnel management processes for entering, developing and training, promoting, and transitioning. These generalized differences are discussed below. The differences are mainly between selected reservists and active duty officers.[1] Active Guard and Reserve (Full-time Support) officers are more like active component officers than like selected reservists. These generalized differences affect the ability of reserve active-status list officers to gain joint experience and education compared to active duty list officers.

---

[1] This section draws heavily from Harry J. Thie, et al., *Factors to Consider in Blending Active and Reserve Manpower Within Military Units,* RAND Corporation, MG-527-OSD, forthcoming.

Entering. This includes entering the component, as well as entering a first or subsequent unit. The active component hires nationally, uses a closed system that prefers those without prior service, provides initial entry and occupational training at entry, and assigns qualified personnel to a duty position in an organization. The reserve component hires locally, uses a more open system that prefers those with prior active component experience, and provides initial entry and occupational training/retraining over a long period of time. Hiring and assigning are largely simultaneous in the reserve component.

Developing. The active component develops human capital through planned horizontal and vertical job rotations (including geographical rotations) and periodic training and education that occur at fixed periods in a career path. It is a time-based system. The reserve component develops human capital through local use (on the job), and episodic training and education as positions become available. It is more of an event-based system. The active component system is a push system; officers are assigned to positions. The reserve component system is a pull system; positions are sought out by officers.

Promoting. The active component system is a rank in person system. Personnel are selected for and promoted against service-wide vacancies and eventually placed in a position at the higher grade. Once achieved, rank is kept regardless of position serving in. The reserve component system is largely (exceptions exist) a rank in job system. A position must be found in order to be promoted; rank is "lost"[2] when the person is no longer in a position for that rank.

Transitioning. Entry into each of the components is described above. Once in a component, the person could move from one component to another or out of the component completely (separation or retirement). The active component uses a time-based system for defined benefit retirement after certain years of service. The reserve compo-

---

[2] Each service manages this somewhat differently. For example, in some, one must find a position in order to be promoted; in others, promotion occurs, but if a position is not found, the officer must move to nonpay status. In general, an officer must find another position, go to the Individual Ready Reserve (IRR), retire, or resign his or her commission.

nent uses a points (event) and time-based system for defined benefit retirement at a certain age. Transition between the two components is administratively complex.

Law, Policy, and Behaviors

In law, the Reserve Officer Personnel Management Act was modeled on the Defense Officer Personnel Management Act that preceded it. As a result, reserve active-status list and active duty list management are a complex interaction of law, policy, and behaviors. Besides the generalized workforce differences, there are specific differences in law and policy that compound the general differences. Moreover, the active duty list officers could be considered as a largely homogenous group to which policy applies similarly. However, the reserve active-status list officers are more heterogeneous in that different policies apply differently to three key groups: unit (or traditional M-day) reservists, Individual Mobilization Augmentees, and Active Guard and Reserve.

Management Frameworks

In previous work,[3] we outlined management frameworks that are conceptual ways to analyze and model officer development, management, and assignment practices. Each of these frameworks leads to differences in assignment tenures, promotion, and retention that must be considered in analyzing supply against demand. There are four frameworks at work within an organization or community. The first three are developmental and management paradigms in that conscious thought goes into a series of horizontal and vertical assignments to develop leaders (managing leader succession), build deep competencies (cadre), or expose officers to new functional areas (managing skills). The fourth is an assignment paradigm in which an available officer is matched to an open position. Our observation is that much of active duty list officer

---

[3] Thie, Harrell, et al., *Framing a Strategic Approach for Joint Officer Management,* RAND Corporation, MG-306-OSD, 2005; Thie, Harrell, and Emmerichs, *Interagency and International Assignments and Officer Career Management,* RAND Corporation, MR-1116-OSD, 1999.

management takes place in the fourth framework and less takes palce in the first three.

Some of these frameworks, however, have less application for reserve active-status list officers. There tends not to be a central assignment function that is consciously rotating officers worldwide to fill open positions. Nor is there a widespread opportunity to move laterally to obtain different functional skills. Some opportunity exists for vertical mobility, but it tends to be as a result of finding local or regional vacancies rather than as part of a managed career. If any one of the paradigms holds for reserve active-status list officers, it is the cadre paradigm where officers stay in jobs and units for long periods of time.

Gaining joint experience and education is difficult for reserve active-status list officers because of limits of time and geography. Most unit-based officers have full-time jobs and serve for at least the minimum required days and drill periods. Time constrains education even when it can be accomplished through distance learning. Geography constrains the availability of positions that give joint experience.[4]

Producing Joint Officers

Active Component

Before describing the current approach to RC JOM, it might be helpful to briefly review how joint officers are produced in the active components. The Goldwater-Nichols Act (GNA) established a new classification of officers—joint specialty officers, or JSOs—who were to be "particularly trained in and oriented toward joint matters." The typical path to becoming a JSO is as follows: An officer attends Joint Professional Military Education (JPME) first, serves a joint duty tour as a JSO nominee, and eventually is designated a JSO. To meet the educational prerequisites to become a JSO/JSO nominee, an officer must, at a minimum, complete one of the following:

---

[4] Changes such as a more "operational" reserve, continuum of service, variable service units, and the Army's force generation model are likely to have effects as well.

- JPME Phase I at Service intermediate or senior-level college (accredited).[5]
- JPME Phase II at the National War College (NWC), Industrial College of the Armed Forces (ICAF), or Joint Forces Staff College (JFSC).

As of next year, senior-level service programs will become eligible for future accreditation for JPME II. These include U.S. Army War College, U.S. Navy College of Naval Warfare, U.S. Marine Corps War College, and U.S. Air Force Air War College.

Credit for joint duty assignments (JDAs) is limited to billets on the Joint Duty Assignment List (JDAL), as mandated by GNA. The list includes those positions at organizations, outside the individual services, that address issues involving multiple services or other nations where the assigned officer gains a "significant experience in joint matters." Thus, to become a JSO, an officer typically needs to have both joint education and prior joint experience, with the sequence being JPME first, followed by a JDA. There are exceptions, such as officers in critical occupational specialties or officers with a waiver by the Secretary of Defense who may serve a joint tour first and then attend JPME. In a few cases, an officer can qualify by completing two joint tours without attending JPME; but again, this requires a waiver by the Secretary of Defense. The number of out-of-sequence and two-tour waivers is limited by law.

Reserve Component

As mentioned earlier in the report, the GNA of 1986 specifically required the Secretary of Defense to establish personnel policies emphasizing education and experience in joint matters for reserve officers on the Reserve Active Status List (RASL)[6] that—to the extent practicable—

---

[5] Officers (other than those with a critical occupational specialty) must attend JPME Phase II before completing their joint assignment to qualify as a JSO. Attendance at JPME II before completing JPME I requires a waiver by CJCS.

[6] The RASL is a single list of officers who are in an active status in a reserve component of the four services and not on the active duty list, and is required by law (Section 14002 of Title 10).

mirrored those for active joint officers. Despite this requirement, little was done to develop and implement policies for reserve joint officer management until recently.

Joint Education for Reserve Active-Status List Officers

In 1995, the Logistics Management Institute (LMI) was asked to determine the need for JPME for reserve component officers.[7] It found that approximately 4,400 reserve component (RC) officers were assigned to joint organizations (defined as those having positions on the JDAL). More than 90 percent of these "joint" officers were Individual Mobilization Augmentees (IMAs), and the others were full-time Active Guard/Reserve (AGR) officers.[8] Even IMAs assigned to joint organizations may be asked to deploy (typically) for peacekeeping or police missions, depending on availability. Thus, there is a distinct need for joint education and training for these officers.

Because most reservists are part time, their educational experiences are largely part time as well. The service intermediate and senior schools offer JPME I embedded in their resident and nonresident program. Although appreciable numbers of reserve officers are able to complete JPME I, this varies by service.

However, few RC officers have the time or the opportunity to complete JPME II. JPME II is offered only through resident programs that range from 12 weeks to 1 year. There are two reasons for this: (a) the learning objectives for Phase II emphasize application of military theory, principles, and practices under joint conditions, and this is achieved through practice and repetition under "real world" (or simulated) conditions; (b) students are steeped in the culture and prac-

---

[7] Dayton S. Pickett, David A. Smith, and Elizabeth B. Dial, *Joint Professional Military Education for Reserve Component Officers,* McLean, VA: Logistics Management Institute, 1998.

[8] IMAs are reservists who attend drills but are preassigned to an active component organization or other federal billet that must be filled on, or shortly after, mobilization. IMAs train on a part-time basis with these organizations to prepare for mobilization. Inactive duty training for IMAs is decided by component policy and can vary from 0 to 48 drills a year. IMAs, by DoDD 1235.11, are not to be assigned to units of the RC force structure. AGRs serve full time in particular assignments to include the RC force structure.

tice of other services and other militaries by working and living with classmates from services other than their own. It would be difficult to fully realize these objectives through nonresident programs. In addition, because of the high need for active component (AC) officers to be trained, very few RC officers are selected to attend these programs (and very few officers can afford the time required to do so, even if selected).

LMI found that a large and growing group of RC officers were doing joint work and working on joint matters and that there was a substantial need for JPME. The study recommended that OSD and the Joint Staff establish an advanced JPME program for RC officers serving or selected to serve in joint organizations.

Since then, there has been some progress in the area of reserve component joint education. For example, in keeping with the LMI recommendation, the 1999 National Defense Authorization Act directed DoD to develop a course similar, although not identical, to JPME II that would prepare reserve component field grade officers for joint duty assignments. Congress further specified that periods of in-house training combined with distance learning curriculum would be the most appropriate format for the prescribed course.[9]

Following this, the Joint Forces Staff College (JFSC) instituted a program, Reserve Component Joint Professional Military Education (RC JPME), tasked with identifying and developing JPME opportunities for RC officers. As its first task, this group developed the Advanced Joint Professional Military Education course, or AJPME. The name served to distinguish it from JFSC's JPME II course. AJPME is modeled on the JFSC Joint and Combined Warfighting School (JCWS), and its curriculum includes National Security Systems; Command Structures; Military Capabilities; Theater (Combatant Command) Campaign Planning with Joint, Multinational, and Interagency Assets; the Joint Operation Planning and Execution System; and Integration of Battlespace Support Systems. As directed by Congress, the course combines in-house education with distance learning. AJPME is a 40-

---

[9] Director, RC JPME, JFSC, Information Paper on Joint Professional Military Education (JPME) for the Reserve Components (RC), dated August 25, 2004, unpublished.

week blended course with 3 weeks of in-residence "face time."[10] The course is considered to meet the education requirement for becoming a fully qualified joint officer or JSO-equivalent (currently reserve officers are not eligible for the JSO designation). AJPME's inaugural class began in the fall of 2003 and graduated the following spring, and about 225 officers have graduated as of January 2006.

One big omission is that personnel database systems have not been updated to include these new AJPME graduates, and no tracking mechanism exists to identify reserve active-status list officers with joint education and/or experience.[11]

In addition to AJPME, reserve officers may take the JPME II course at the senior level service schools. These will become accredited for JPME II this year, and there are a small number of JPME II seats (66) reserved for RC officers: U.S. Army War College (45 seats); U.S. Air Force Air War College (20 seats); U.S. Marine Corps War College (1 seat). As of now, there are no seats reserved for RC officers at the U.S. Navy College of Naval Warfare.

To summarize, currently the education requirements for RC joint officers are as outlined in Department of Defense Instruction (DoDI) 1215.20, issued in 2002:

- To the maximum extent practicable, officers on the reserve active-status list should complete the Officer Professional Military Education Policy (OPMEP) outlined in the Chairman of the Joint Chiefs of Staff Instruction 1800.01A issued in 2000 and JPME Phase I before being assigned to a joint duty assignment that requires basic JPME. Officers who are to be assigned to critical billets identified as requiring advanced JPME should complete advanced JPME (to the maximum extent practicable) before the assignment.

---

[10] LTC Judith A. Davenport, "An Analysis of Reserve Component Joint Officer Management Including Five Major Issues and Suggested Recommendations," dated March 13, 2004, unpublished.

[11] Davenport, ibid.

- Completion of JPME Phase I is required before enrollment in advanced JPME.

Joint Duty Assignments for Reserve Active-Status List Officers

DoD made little headway in identifying a joint duty assignment list for RC officers that would qualify them to receive joint duty credit. As we saw above, even in the mid-1990s, well over 4,000 RC officers were serving in joint organizations, and many were working on joint matters, although they received no formal credit for such assignments. Indeed, in 2000, a study by the Congressional Research Service concluded that the DoD had made no progress in meeting the GNA requirement and that the reserve components lacked procedures to identify and track positions that provided a valid joint experience.[12] In 2002, DoD attempted to concretize roles and responsibilities for RC JOM by issuing DoDI 1215.20. Its purpose was to "implement policy, assign responsibilities, and prescribe procedures for administering joint officer management for officers on the DoD Reserve Active Status List (RASL)."[13]

The Assistant Secretary of Defense for Reserve Affairs was given oversight responsibility of RC joint officer management and directed to coordinate RC joint officer education and joint officer management issues with the Chairman of the Joint Chiefs of Staff (CJCS). The latter had the responsibility of establishing education and examination criteria and of validating and documenting Joint Duty Assignments-Reserve (JDA-R) in officers and agencies. The validation was to include identification of JDA-R that require no JPME, those that require JPME Phase I, and critical billets that require advanced JPME. The Secretary of Military Departments and the Commandant of the Coast Guard

---

[12] Congressional Research Service, *Department of Defense Reorganization Act of 1986: Proposals for Reforming the Joint Officer Personnel Management Program*, Washington D.C., July 18, 2000.

[13] Department of Defense Instruction, *Reserve Component (RC) Joint Officer Management Program*, Number 1215.20, September 12, 2002.

were given the responsibility of facilitating RC officer JPME and validating and documenting JDA-R in their respective military services, applicable defense agencies, and non-DoD entities.

The DoDI also outlined rules and regulations with respect to JPME requirements, assignments, and what should be included in JDA-R. We described the JPME requirements earlier. Here we turn to (a) the rules governing assignments and (b) the guidelines regarding JDA-R.

Assignments. DoDI 1215.20 states that the graduates of the National Defense University (JFSC, ICAF, and NWC) who are on the RASL should be assigned to a JDA-R within 3 years of successful completion of the course (to the extent feasible). In addition, minimum tour lengths are defined as 2 years for officers in full-time support status and 3 years for officers not in full-time support status. In special circumstances (selection for senior military service school or selection for a command assignment that cannot be delayed), the tour lengths could be shortened to 18 and 30 months, respectively. However, credit for designation as a fully joint qualified officer depends on completion of cumulative assignment requirements.

JDA-R Categories. DoDI 1215.20 provides the following general guidelines for determining if positions may be classified as JDA-R:

6.4.1. *OSD Positions.* All positions on the OSD staff where the incumbents shall be responsible for developing and promulgating policies in support of U.S. security objectives. Positions on the Reserve Forces Policy Board shall be included in this category.

6.4.2. *Joint Staff Positions.* All positions where the incumbents shall be involved in the national military strategy (NMS), joint doctrine, education, training, policy, strategic planning, or contingency planning.

6.4.3. *Combatant Command Positions.* All positions where the incumbents shall be involved in the NMS, joint training and exercises, strategic planning, contingency planning, managing resources, and command and control of combat operations under a Combatant Command.

6.4.4. *Organizational Positions (Other Than Those in the OSD, the Joint Staff, or the Combatant Commands).* The incumbents of those positions shall be involved with the integrated employment or support of land, sea, and air forces, of at least two Military Services. Most of their duties shall deal directly with matters relating to the NMS, joint doctrine or policy, strategic planning, contingency planning, or command and control of operations in support of a Combatant Command.[14]

Current Status of JDA-R. Despite the requirement to identify billets that would qualify as JDA-R, little progress has been made in this area. Several examples of the less than optimal nature of the current state of affairs abound. No joint duty credit accrues to RC members serving at the Office of the Secretary of Defense, Joint Chiefs of Staff, Unified Combatant Commands, and Joint Task Force headquarters or in various other joint centers (for example, the 27 Joint Reserve Intelligence Centers established under the Joint Reserve Intelligence Program set up to assist the defense intelligence community in meeting intelligence missions). Further, the Reserve Forces Policy Board (RFPB) (2004) points out that some of the RC personnel (especially RC full-time support personnel) are assigned to billets that are considered liaison positions within the organizations and thus are not even included in the organization manning/authorization document.

As the RFPB points out:

> Full integration of the RCs in Joint Operations is no longer an idea, but a reality of how business is being accomplished . . . With the new steady and future state of increased RC involvement, training, equipping, maintaining, and educating our members to a similar level of our active duty counterparts is reality. (p. 27)

The next chapter outlines a possible strategic approach to coming closer to that reality.

---

[14] DoD Instruction, pp. 5–6.

Assessment

In this conceptual phase, our intent is to discuss, understand, and reconcile the existing approaches for crediting or qualifying jointness in the officer corps. This section provides our assessment of lexicon and measurement and offers suggestions for reconciling the first and improving the second.

Definitional Differences

As stated above, DoD Instruction 1300.20 of 1996[15] conforms to directions set forth in the GNA for active duty list officers. A joint duty assignment is defined as one in which the officer gains significant experience in joint matters and a list of such assignments is maintained. Part of the definition includes an association with officers of at least two of the three Military Departments. Moreover, if an officer gains education and experience of particular types for particular durations, he becomes a Joint Specialty Officer, or someone who is trained in and oriented toward joint matters. If an officer has completed JPME II but not a JDA, he is considered a JSO nomination.[16]

The DoD instruction on RC joint officer management issued in 2002 introduces two additional terms to the lexicon, *Fully Joint Qualified* and *Joint Officer*. The first requires AJPME and sufficient time in a qualifying billet. This is not unlike a JSO. The second term defines an officer who has achieved CJCS joint learning objectives[17] and has served or shall serve in a JDA-R billet. For one who has served in a JDA-R billet, the difference between being a Joint Officer and Fully Joint Qualified appears to be in achieving CJCS learning objectives through a process other than AJPME.

---

[15] DoD Directive 1300.19, as of November 2003, is the policy document and the instruction provides specific guidance. The directive was first issued in 1978, predating GNA, and updated several times.

[16] An officer in a designated critical occupational specialty who completes a joint duty assignment is also considered to be a JSO nominee.

[17] The instruction does not define how this is measured. It could be through school attendance or self-study or other means.

Most recently, the Chairman of the Joint Chiefs of Staff issued his vision of joint officer development.[18] JOD uses the terms *fully qualified* and *joint qualified* as synonyms and concludes that the term *Joint Qualified Officer* (JQO) should replace the term *JSO*. The component pieces of JQO are joint individual training (JIT), JPME I and II, and experience, all of which could be obtained through multiple paths. A general or flag officer would have to certify performance for all qualifications even in the traditional time and billet system.

Are these differences reconcilable? Yes. We offer observations here and further refine them in Chapter Five. Fully Joint Qualified in the DoD RC JOM instruction is equivalent to JSO in the AC instruction. Both are achieved through an advanced JPME education seat and an assignment to a specified joint billet for a specified time. A JQO as outlined in the CJCS JOD (we assume it is meant to apply to both active and reserve) includes the above as a minimum means (education- and billet-based time system) but also includes officers who qualify on other paths.[19] For example, a Joint Officer (RC) could demonstrate achievement of CJCS learning objectives gained through multiple paths (including experience of less than the specified time). Once such an officer serves in a specified JDA-R billet[20] for the specified time, the officer would be Fully Joint Qualified and a "Joint Qualified Officer."

Thus it appears that a reserve active-status lexicon can reconcile the DoD Instruction for RC JOM and the CJCS JOD by using a two-level qualification approach. The first level is "Joint Officer" and the second level is "Joint Qualified Officer." Either could be achieved by the traditional time- and billet-based system or by other means. To be a Joint Officer requires JPME I and II/AJPME or achievement of learning objectives, with or without some experience. To be a Joint Qualified Officer requires being a Joint Officer and serving in a specified billet for requisite time or having greater degrees of experience certified by some other means. Ultimately, the requirement boils down to knowledge

---

[18] *CJCS Vision for Joint Officer Development*, November 2005.

[19] We will suggest such paths for reserve active-status officers in Chapter Five, and those paths might become the basis for changing the constraints now in law for active duty officers.

[20] Later, we address whether serving in a JDAL billet should also count.

or experience for the first level and knowledge and experience for the second level with joint acculturation underlying both. This is not unlike the current JSO nomination and JSO system for active duty officers, so we believe that reconciliation of the lexicon among all is achievable.

We use the *Joint Officer* and *Joint Qualified Officer* terms in this report and further discuss them in Chapter Five.

Performance Measurement

For the conceptual approach to joint officer development for reserve active-status list officers, we also assert a more modern approach to performance management.

Systems can be viewed as inputs, processes or activities, outputs, and outcomes. Currently, the active duty JDAL approach is to measure and validate the activities/processes under the assumption that they achieve the desired outputs. This is called a process evaluation, and the goal is to assess the extent to which a program is operating as it was intended to operate. "It typically assesses program activities' conformance to statutory and regulatory requirements . . . and . . . expectations."[21] So, for example, the current active duty approach to joint performance measurement is to determine carefully which assignments qualify; the duration for which particular officers must serve in those assignments; whether the education curriculum is tied to missions, capabilities, and competencies, whether the classroom has sufficient "associating" for full acculturation; and if promotions are within bounds. It is process control.

We are asserting that there are other activities and processes for developing officers in jointness (not all visible) beyond the formal ones used by the active duty system and that it may be more useful to include measuring the outputs/outcomes as part of the performance measurement and evaluation system. This outcome evaluation "focuses on outputs and outcomes . . . to judge program effectiveness but may also assess program process to understand how outcomes are produced."[22]

---

[21] General Accountability Office, *Performance Measurement and Evaluation: Definitions and Relationships* (GAO-05-739SP), May 2005.

[22] Ibid.

Figure 2.1 depicts the two kinds of performance measurement. Ultimately, we suggest allowing both kinds of measurement and validation. In essence, a process can be measured for conformance as an indicator of output and outcome, or officers can be assessed directly by other means.

Figure 2.1
Performance Measurement

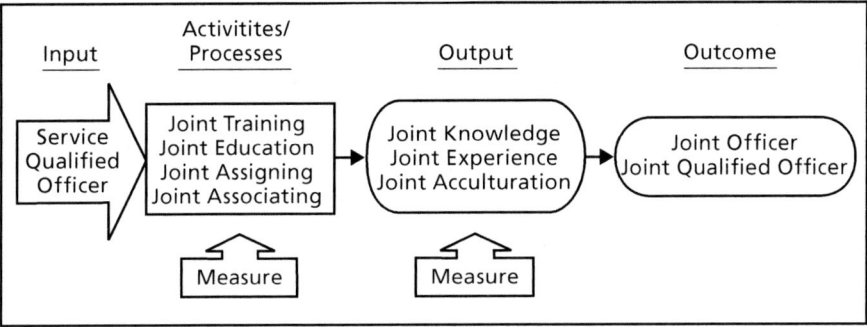

CHAPTER THREE

A Strategic Approach to Joint Officer Management: Why and How

This chapter provides a rationale for a strategic approach to joint officer management and uses our work on active component officers as an example to illustrate the need for and how to implement such an approach. The chapter also discusses some of our continuing work in this area, which will provide valuable lessons learned regarding developing and implementing a strategic approach to Reserve Component Joint Officer Management.

Need for a Strategic Approach to Joint Officer Management

The President's National Security Strategy as well as service and joint vision documents describes increasingly joint missions. Our previous study of active duty list officers shows that officer management is following the trend more slowly.[1] Data indicate a decreasing fill rate of Joint Duty Assignments (JDAs) for three of the four services. (The Marine Corps is not only increasing the rate at which it fills joint assignments but also increasing its share of the Joint Duty Assignment List (JDAL).) The service personnel managers (for all but the Marine Corps) note the difficulty in fitting joint assignments into officers' career paths and are reflective of individual service cultures that are generally less respectful of joint experience than of that gained within their services.

---

[1] Thie and Harrell, et al., ibid.

Nonetheless, longitudinal data indicate an increasing amount of jointness among the officer corps, although this is more true for certain grades, occupations, and services than for others. A cursory look at the data, however, indicates that such increases are leveling off, that is, becoming asymptotic at current levels. How joint the officer corps can be is dependent on the opportunity to have a joint duty assignment and to attend the second phase of the Joint Professional Military Education (JPME II). The seats for the latter are limited, and the number of the former is also limited. It may be that given these constraints, underlying job and educational durations, and continuation and promotion rates, the ability to increase jointness further in the officer corps in the future may not exist absent changes in the number and duration of school and assignment seats.

A strategic approach must understand the need or requirement for critical workforce characteristics and the ability of the management system to provide officers with those characteristics. Moreover, the approach needs to demonstrate (a) a strategy or policy for aligning the availability of officers who have the characteristics with the need for them, or (b) a rationale for why more widespread availability of the characteristics rather than the immediate need for them would be desirable.

Why a Strategic Approach? A Conceptual Framework[2]
Apart from how to take a strategic approach, why should an organization take a strategic approach? The basic reasons are that (a) human resources management in an organization has multiple influences that are often in conflict, and (b) primary actors within an organization with different values and attitudes have leeway to make human resources choices. A strategic approach determines which influences are more important and limits choices of the actors to those most conducive to organizational performance.

---

[2] This section is based on Paauwe and Boselie and authors cited by them. Paauwe, J. & P. Boselie, *Challenging (Strategic) Human Resource Management Theory: Integration of Resource-Based Approaches and New Institutionalism*, Erasmus Institute of Management (ERS-2002–40-ORG), Rotterdam, The Netherlands, April 2002.

Influences

Three organizational influences affect human resources management. The first is the organization's administrative heritage. This influence is that of structures, methods, and competencies that originated in the past. For the military it is the lingering legacies of a Department of the Navy and a Department of War. The military services have long memories with respect to how human resources management has been done. The services know how to develop, educate and train, assign, and promote officers. They have been doing it for two centuries and have been operating the basic design of their systems since at least the end of WW II with the passage of the Officer Personnel Act of 1947. The Grade Limitation Act of 1955 and the Defense Officer Personnel Management Act of 1980 extended these basic designs, and the Reserve Officer Personnel Management Act was modeled on the Defense Officer Personnel Management Act (DOPMA). The Goldwater-Nichols Act (GNA) disrupts them by mandating that different career paths and developmental practices are needed.

The second influence is the cultural and legal influence, which imposes currently prevailing values and norms such as fairness, equity, merit, and equality of opportunity to establish relationships with both internal and external stakeholders. Officership is the issue, and it has been service based. The military operates within a closed system and a fundamentally different legal paradigm than that of the private sector; notions of fairness, equity, and merit are different between the two. Age and other discriminations are practiced in one in ways that are illegal in the other. The Fair Labor Standards Act and the Employee Retirement and Income Security Act are among the nation's laws that do not apply. The military has its own legal and social practices as a result of law, executive order, or policy, and there are different legal and social practices for reserve active-status list officers. The GNA is disruptive in that it suggests that some officers are more valued than others in a way that the current system has not accepted.

The third influence is the mission and technology orientation by which national security is produced and delivered. The issues here are efficiency, effectiveness, flexibility, quality, and innovativeness. The GNA disrupted this in significant ways by bringing combatant commanders and their needs to the forefront. The services and DoD have adjusted to this and appear to have even embraced it on the operational side but not on the management side.

Constraining Action While Providing Flexibility

There are multiple decisionmakers in the human resources system. Among them are the Congress, agencies in the executive branch, the Secretary of Defense and his staff, the Chairman of the Joint Chiefs of Staff, the military departments and the chiefs of service, the reserve component chiefs and their respective staffs, organizations that use officers, and the officers themselves. Constraints are needed on the choices that these decisionmakers could make within an unconstrained system; however, the system should not be overly constrained such that it becomes inflexible. Decisions to be made are about "fit." The human resources system must have strategic fit in that it supports the strategy of the military and of its organizations. It must be based in the need to be successful in prosecuting military operations and delivering national security. The system must also have organizational fit in that the human resources (HR) system must work in conjunction with other organizational and administrative systems such as the deployment and readiness systems, which are themselves changing. The HR system must have environmental fit. The strategies used must be in consonance with the practices and norms of the larger external environment and with the needs of prospective and serving officers. And last, the system must have internal fit. HR practices must be coherent and consistent bundles of policies and practices.

Shaping HR Strategies Toward Organizational Performance

Military operations and organizational fitness for them require shaping HR strategies to generate HR outcomes that contribute to performance of the organization. In essence, the mission and goals influence is emphasized while other influences such as administrative heritage and cultural norms are recognized. By suggesting a new competitive strategy for the military that leads to changes in organizational and administrative systems and cultures, GNA has imposed different constraints on the multiple decisionmakers in the system that require adjustments. A strategic approach to human resources management is important to achieving organizational goals and missions.

Framing a Strategic Approach

A strategic approach for joint officer management must deliberately determine which jobs, inside or outside the service, require joint experience or provide it. In particular, given the current strategic intent of the Department with respect to jointness ("push it to its lowest appropriate level"), the need for joint experience should be measurable in a much larger number of billets, in particular in billets internal to the service. Moreover, valid joint experience might now be provided by service in billets internal to the service, particularly those associated with Joint Task Forces, with service component commands, and with joint planning and operations. The key components of a strategic approach can be discerned as (a) which jobs require or provide joint experience, (b) how many of each exist, and (c) what is needed to align those two sets of jobs.

Workforce Characteristics

There are three well-known requirements in law for active duty list officers from which we can infer need for one or the other of two critical workforce characteristics: joint experience and joint education. First, the requirement for active duty list officers to have completed a JDA before promotion to general or flag rank sets a requirement for joint experience for most of the approximately 900 active component general and flag officer (GFO) positions.[3] There is also a requirement to fill 800 critical positions with joint specialty officers (JSOs) that sets a requirement for active duty list officers in these positions to have successfully completed JPME II and a prior JDAL assignment. Third, the requirement to fill at least half of JDAL positions with a JSO or JSO nominee sets a requirement for 50+ percent of JDAL positions to be filled with active duty list officers who have completed JPME II. On

---

[3] Some officers, such as doctors, are exempt from this requirement.

the other side, there are constraints either in law or in DoD policy that affect the availability of officers with the joint characteristics. For example, qualifying joint experience can be obtained only in billets external to the military service. These billets for obtaining qualification are further limited in that they must be in the grade cf O-4 and above, and only some of the billets in defense agencies can provide the qualification.

Need for and Availability of Characteristics
The following three notional diagrams portray the contrast for need and availability between the current system and a system premised on a strategic approach. In the current system for active component officers, as shown in Figure 3.1, those billets that have a prerequisite need for joint experience are largely a subset of those that provide joint experience. (Diagrams that portray the need for JPME II would be similar.) Much of the emphasis of Goldwater-Nichols and the DoD implementation has been on identifying active component positions that provide a valid joint experience given that officers serve in them for a minimum amount of time. All such positions (identified by the large circle) must

Figure 3.1
In Current Active Component System, Need for Joint Experience Is a Subset of Availability

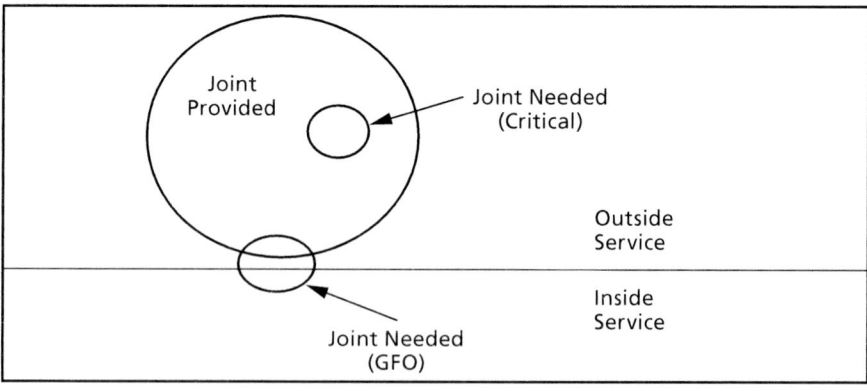

be external to or outside the military service. The need for officers with such experience is the key aspect of a strategic approach and is shown by the two small circles. Joint experience is needed in 800 critical billets that are all outside the service and in most of the 900 general and flag officer billets, many of which are internal to the service.

The current system for reserve active-status list officers, shown in Figure 3.2, is a variation on this diagram. The circle representing those billets that provide joint experience is represented by a dashed line. Although there is a DoD instruction that joint duty assignments in the reserve component should be tracked, no billets have been identified or entered into a formal list. Thus there are no billets currently recognized to provide joint experience. Further, the only need for reserve active-status list officers to have had prior joint duty experience is for the reserve component chiefs. Thus the positions where joint experience is needed are reflected by the small circle. However, that circle is portrayed with a grey line because the requirement for reserve component chiefs to have had prior joint experience may be waived by the Secretary of Defense.[4] Thus there is no proactive recognition or track-

Figure 3.2
Current Reserve Component System Does Not Acknowledge the Need for or the Provision of Joint Experience

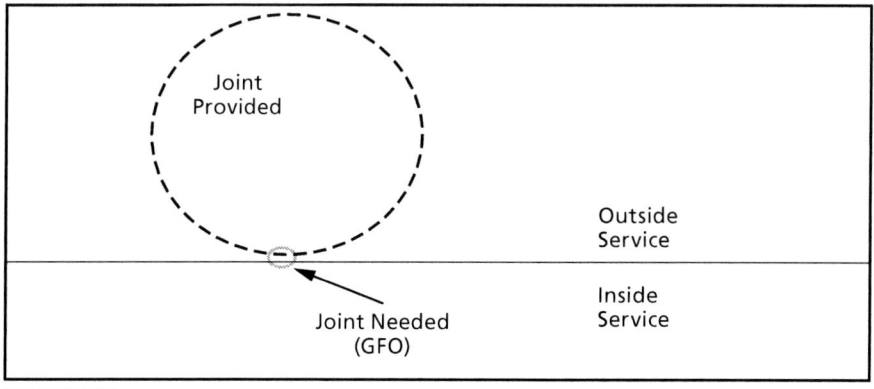

---

[4] The 2005 NDAA extends this waiver authority for two years, to December 31, 2006.

ing of billets for reserve active-status list officers that provide a joint experience, and there is not an active enforcement or recognition of reservist billets that require a joint experience.

In contrast, a strategic approach involves a deliberate recognition and determination of which jobs, inside or outside the service, need joint experience or provide it. Such an approach is portrayed in Figure 3.3. In particular, given the current strategic intent of the Department with respect to jointness ("push it to its lowest appropriate level"), the need for joint experience should be measurable in a much larger number of billets, in particular in billets internal to the service. Moreover, valid joint experience might now be provided by service in billets internal to the service, particularly those associated with Joint Task Forces, with service component commands, and with joint planning and operations. The key components of a strategic approach can be discerned from Figure 3.3—which jobs require or provide joint experience, how many of each exist (what is the size of the two circles), and what is needed to align the two circles.

Figure 3.3
In Strategic Approach, Need for and Availability of Joint Experience Is Determined

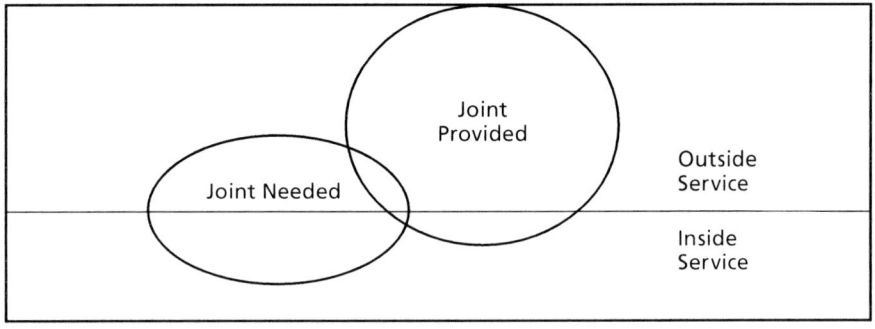

Implementing a Strategic Approach for Reserve Active-Status List Joint Officer Management

Our recommended approach has five major steps:

1. Define workforce characteristics that will be needed in the future to meet strategic intent. We believe that these characteristics can be aggregated into proxy variables for competencies based on experiences such as joint multiservice, joint interagency, joint multinational, and on joint education and/or joint training. The accuracy of billet needs with respect to characteristics such as grade (experience), occupation, and other characteristics will need to be assumed.

2. Define needs for these characteristics of joint experience, education, and training. Where (in what positions) are officers with joint experience, education, and training needed? How many of these positions are there? Does this differ across services, for different occupations, or at different levels of seniority? Does the need for such officers extend to in-service billets?

3. Identify officers with these characteristics who are currently available.

4. Use models to

    a. Project availability of officers with these characteristics in the future, given certain career management practices

    b. Calculate future gaps between the need for officers and the availability of them[5]

    c. Refine and evaluate near-term policy alternatives to reduce gaps within the strategic context

    d. Develop strategies that address long-term issues for reducing the gaps.

---

[5] We are using the logic that underlies strategic human capital management of matching availability of workforce characteristics to the demand for them. This assumes that there is a cost for developing people with these characteristics so that both an oversupply and undersupply of the characteristic is not desirable. However, other assumptions could be made that change the nature of the assessments we are making. For example, the availability of officers with joint experience and/or education could lead to increasing demand for them in many military positions. The availability of such officers could, by itself, create a need for them.

5. Identify other implications of the strategic approach such as effects on objectives and desired metrics for evaluation.

A strategic approach to joint officer management, as outlined here, aligns human capital with the organization's mission, rather than empowering other influences, such as organizational, administrative, and cultural heritage or the current social, cultural, and legal practices and beliefs. The strategic approach described herein for joint officer management considers and balances the assignments that require joint experience, education, training, or acculturation with the ways officers receive joint experience, education, training, or acculturation.

Continuing Work

One of the recommendations that came out of the work on active joint officer management was to undertake a large data collection effort on joint and potentially joint billets to help operationalize the approach outlined above. The 2005 JOM survey was a Web-based survey designed to elicit information on joint billets on the JDAL and potential joint billets in external organizations and internal service billets nominated by the services as requiring or providing joint experience.[6] More than 21,000 officers responded to the survey and provided information on the billets and the extent to which their assignments provided them with joint experience or required them to have had prior joint education, training, or experience. Nearly 700 of them were reserve active-status list officers. We use these responses to illustrate how these data can inform active duty list and reserve active-status list joint officer management policy.

Our strategic approach emphasizes the importance of understanding the demand for and supply of reserve active-status list officers with joint education and/or experience. The next two chapters focus on these two topics.

---

[6] For a more detailed discussion of the 2005 JOM Survey and its results, see Kirby et al. (2006).

CHAPTER FOUR

Determining Demand for Reserve Active-Status List Officers with Prior Joint Education and/or Experience

A key component of developing a strategic approach to joint officer management for reserve active-status list officers is understanding the need for joint education and/or experience in billets for these officers. The current level of utilization of the reserve component and new missions, organizations, and structures stood up within the reserve component (RC) point to an increased recognition that the work required of reservists is becoming increasingly joint.[1] As the Reserve Forces Policy Board (2004) stated:

> Today, the Reserve Components (RCs) are being utilized to a degree to which they haven't experienced since Desert Shield and Desert Storm. The RCs are intimately integrated into the Homeland Defense mission, as well as the entire expeditionary mission, as the Global War on Terror (GWOT) is executed. Joint Operations and the RCs are now full partners, hand in hand, as they work to plan, organize, and equip themselves to fight the wars that lie in front of them. Full integration of the RCs in Joint Operations is no longer an idea, but a reality of how business is being accomplished. Given our current strategic situation, National Security policy and future commitments, the future uti-

---

[1] This is a result of several factors to include more military work being defined as joint, more service work being done in a joint operating environment, and reservists participating in such work and in greater numbers. Moreover, homeland security and homeland defense (HLS/HLD) tasks are increasing, especially with respect to interagency and intergovernmental needs.

lization of the RCs will most likely resemble how it is being used today—with the exception of it being more "Jointly" orientated and tasked. (2004: 27)

Although it is a truism that all forces are jointly integrated and employed by combatant commanders, we are addressing individual reserve active-status list officers. This chapter first examines the potential demand for prior joint education and/or experience in reserve active-status list joint officers, focusing on various developments in the structure of reserve components that are likely to generate such demand. It discusses a one-time data collection exercise that would help provide a foundation for demand. Such information would need to be validated on an ongoing basis, as the duties and functions of the billets change or new billets are added and collected in permanent manpower databases.

Demand for Reserve Active-Status List Officers with Prior Joint Education and/or Experience

Demand for prior jointness is likely to be very different across the different reserve categories. For example, the typical reserve unit staffed by drilling reservists is unlikely to generate a demand for officers educated and trained in joint matters, with perhaps a few exceptions. More than 70 percent of O-4 and above selected reservists are in this category. The exceptions could be unit reservists who may volunteer for active duty for special work, for example, in a Joint Task Force (JTF). However, the number that volunteer goes up and down with the standup of JTF that cannot be sourced internally by the combatant command or with active duty list officers. In addition, as we mentioned earlier, individual mobilization augmentees (IMAs) are generally assigned to active units, and some of these positions are in joint organizations. This is also true of active guard reserves (AGRs) who are serving in external organization billets, for example, at the Office of the Secretary of Defense (OSD) or on the Joint Staff, or in combatant commands. So we would expect that much of the demand for IMAs and AGRs with joint experience or education would be a derived demand from active component organizations, external to the military service, with little demand within service reserve component units.

Demand for prior jointness is also likely to be very different across the various reserve components. For example, Navy reserve officers typically serve as IMAs in active component organizations or in unit augmentations to Navy active component units. However, National Guard units have both federal and state missions and may be called up to coordinate homeland defense or disaster relief with a number of state and federal agencies. We expand on this point below when we discuss the new joint headquarters being established by the National Guard Bureau. It is very likely that officers in or leading these units would benefit from prior education and experience in joint matters, especially interagency and intergovernmental matters. Moreover, IMAs are not provided by the Air National Guard and the Army National Guard, so this category of demand does not exist. It is primarily AGR and some unit and headquarters' demand for the National Guard.

Demand for prior jointness is also likely to be very different across the various kinds of jointness. For active duty list officers, the greatest demand is for multiservice qualified officers, then multinational, and last interagency qualified officers. For the reserve components, especially the Army National Guard and the Air National Guard, interagency qualifications should be in far greater demand than multinational qualifications. Moreover, the need for intergovernmental experience (state and local) also plays a role in demand that has not been seen for active duty list officers.

Department of Defense Instruction on Joint Duty Assignment-Reserve Categories

The DoD Instruction referenced earlier contained an organizational grouping for determining if billets should be represented on the JDA-R, and the aggregated organizations appear useful also for determining demand. OSD positions, Joint Staff positions, and combatant command positions are primarily examples of active component organizations that use reserve active-status list officers. Organizational positions (other than those in the OSD, the Joint Staff, or the Combatant Commands) would then include such organizations as defense agencies, service headquarters and component commands, and reserve component organizations.

That reserve personnel are working on joint matters and in positions where applicable joint training or education might be desirable is amply demonstrated by an earlier LMI study that examined close to 4,400 RC officer billets, grades O-4 through O-6, in joint organizations.[2] Of these, about 2,500 positions were with the combatant commands, 240 were with OSD and the Joint Staff, and another 1,700 were with defense agencies and activities, including more than 800 with Defense Intelligence Agency and 750 with Defense Logistics Agency. The vast majority (more than 90 percent) of these billets were filled by IMAs, the remainder by AGRs. They surveyed the active component or civilian supervisors of the billets about the duties and responsibilities of the position and the need for basic and advanced Joint Professional Military Education (JPME). With this information, LMI estimated that (a) about 1,900 positions, or 45 percent of all the RC officer positions authorized, required basic JPME and (b) about 1,200 positions, or 28 percent, required advanced JPME. Two things are evident from this study: RC officers are indeed working on joint matters, and the need for joint education far outstrips the supply of RC officers with the requisite education in joint matters. As of the end of January 2006, 225 officers had graduated from Advanced Joint Professional Military Education (AJPME).

Examples of Joint Reserve Component Programs Where Demand for Prior Joint Education and/or Experience Might Exist

In addition to OSD and Joint Staff (JS) positions where such demand should exist, we provide various examples of joint RC units or programs that are likely to require that officers serving in these positions be experienced and educated in joint matters. This list is not comprehensive.

Joint Reserve Units. U.S. Joint Forces Command (JFCOM) is tasked with the transformation of the military and coordination of the provision of resources to U.S. commanders around the world. The JS approved the Joint Reserve Unit (JRU) concept in 1995 and the first

---

[2] Pickett, Smith, & Dial, 1998, op cit.

formal unit was formed in 1996. The JRU is a command and control organization responsible for subordinate reserve units[3] that incorporate all seven of the armed service reserve components (Army, Navy, Air Force, Marine Corps, Army National Guard, Air National Guard, and Coast Guard Reserve). The JRU consolidates administrative, training, and security functions common to all reserve component service elements. Whereas the JRU headquarters performs administrative rather than national military strategy (NMS) duties, the subordinate reserve units serve within JFCOM. Anecdotally, at least some of the unit members should have prior joint education and experience before assignment.

Joint Reserve Directorate. The concept of the Joint Reserve Directorate (JRD) was developed before September 11, 2001, and quickly implemented soon after. The JRD functions as a bridge between the active and reserve units—a one-stop shop through which reserve expertise can be quickly and efficiently accessed, and it is the only reserve directorate at the unified command level. As of 2005, the JFCOM joint reserve directorate numbered over 1,300 reservists. Officers serving in these positions are clearly working on multiservice, interagency, and very likely multinational matters—the very definition of joint assignments. The demand question is how many of them should have prior joint experience or education or both.

Standing Joint Force Headquarters. All of the combatant commands are standing up these organizations as planning cells that can be the nucleus of a JTF when needed. JFCOM in particular has created positions in its Standing Joint Force Headquarters (SJFHQ) that in the future would be filled with reserve active-status list officers. One of the limitations in doing this is in finding sufficiently educated and experienced officers to fill the positions. An interesting question raised is whether a larger supply of qualified officers would increase the demand for the positions, for example, in other combatant commands. That is, would combatant commands and other organizations make more use of reserve active-status list officers if more officers who were

---

[3] These are primarily troop program units for drilling reservists. Some services include an IMA detachment as well.

joint qualified were available? The JS is studying the role that reserve active-status list officers should fill in both rapid deployment and longer term sustainment of Joint Task Force Headquarters (JTF HQS).

United States Northern Command (NORTHCOM). NORTHCOM was established in February 2002 to provide command and control for DoD's homeland defense efforts and to coordinate efforts with civil agencies. NORTHCOM is not considered a "first responder"; therefore, it has few permanently assigned forces. However, its very function requires it to support other federal agencies (unless DoD is asked to take on the lead role) and to work with a variety of federal, state, nongovernmental, and foreign agencies. Some reserve active-status list officers assigned to NORTHCOM or to its service components or to its JTFs are likely to need both experience and education in joint matters.

Joint Reserve Intelligence Program. The Joint Reserve Intelligence Program was established in 1995 by a DoD memorandum, "Peacetime Use of Reserve Component Intelligence Elements," issued by the Deputy Secretary of Defense. As the Reserve Forces Policy Board (RFPB) (2004) report points out, "[T]he plan's vision was revolutionary in that it directed the defense intelligence community to train reservists for mobilization by engaging them in "real-world" mission during peacetime" (p. 28).

There are 27 Joint Reserve Intelligence Centers located throughout the continental United States (CONUS), and each of them is tasked with providing resources and reserve support to the defense intelligence community. The Program has funds to allow the Joint Military Intelligence Program to use RC intelligence elements for intelligence operations, training, and support to meet critical needs of the unified combatant commands. Thus the AC members can quickly and reliably task RC personnel with contingency, crisis, or peacetime requirements, and the Joint Reserve Intelligence Connectivity Program ensures that the various participants are linked in a virtual environment. During times of high operations tempo, however, they tend to get tasked by their own service and are not available to other agencies or combatant commands, suggesting less need for prior joint education and/or experience.

National Guard Bureau. In 2003, the National Guard transformed itself to increase its ability to work in a joint environment and to provide enhanced accessibility to its members and assets. Its transformation plan explicitly states that one of its objectives is to ensure that all personnel in the National Guard are trained to operate in a joint environment. The National Guard Bureau reorganized itself into a joint organization and consolidated its 162 State headquarters organizations into 54 doctrinally aligned Joint Task Force Headquarters that have oversight and responsibility for all Army and Air Guard activities in each state. The Bureau is working to obtain JS approval for the integration of the headquarters organization into the joint manpower process. Thus the Bureau is seeking recognition, in both law and policy, as a joint activity of the DoD and a joint bureau of the Departments of the Army and the Air Force. In addition, it is seeking credit for performance of joint duty and asking for Joint Specialty Officer billets to be allotted to the Guard, as well as increased access to the second phase of the Joint Professional Military Education (JPME II). Another of the initiatives launched by the National Guard is to institute a reserve officer exchange program in which Navy and Marine Corps Reserve officers serve as part of the Bureau staff, and Guard officers, in turn, are assigned to their staffs (2004: 42).[4]

Certainly, many of the assignments outlined above would need to be staffed with officers with prior joint education and experience. How many of these is not known. As with the active component, no database contains such demand information.

Findings on Reserve Active-Status List Officers from the 2005 Joint Officer Management Survey[5]

Among the respondents, 679 were reserve active-status list officer incumbents of positions. One caveat is unique to the reserve data shown here. The survey was targeted at active component billets, and

---

[4] The National Guard Bureau, *National Guard 2005 Posture Statement,* March 2004.

[5] For more detailed findings on all officers, see Kirby et al. (2006).

the fact that 679 reservists and guardsmen who were billet incumbents responded is simply serendipitous. This raises two concerns about the representativeness of these respondents:

1.   They were in the survey because they were assigned to billets that were on the joint duty assignment list (JDAL), in external organizations, or explicitly nominated by the services as potentially joint. Thus the information we have is about these specific billets and says nothing about the nature—joint or otherwise—of other reserve billets.
2.   Although the survey did not collect information on the reserve status of the respondents, it is likely that they were AGRs or IMAs. Thus, at best, their experiences and opinions will be similar to those of other AGRs and IMAs serving in external organizations and will not represent those of unit reservists.

Findings

Of the 679 reserve active-status list officers who were billet incumbents, 69 (10 percent) were serving in JDAL billets, 431 (64 percent) were in external organization billets, and 179 (26 percent) were serving in service-nominated billets. Somewhat surprisingly, 150 of these officers were junior officers, 19 were general and flag officers, and the remainder (n = 510) were mid-grade officers.

In this chapter and the next, we focus only on officers in grades O-4–O-6. Of the 510 reserve officers, 35 percent were O-4s, 41 percent were O-5s, and 24 percent were O-6s. Figure 4.1 shows the distribution of these officers by reserve component. The bulk were in the Army Reserve, while the Army National Guard, the Naval Reserve, the Air Force Reserve, and the Marine Corps Reserve accounted for between 10 and 16 percent of the officer incumbents. Only 14 of the 510 reserve officers were Air National Guard officers.

Table 4.1 shows the major billet organization where these officers were assigned. Close to 70 percent were in external organization billets, and 19 percent were in service-nominated billets. A small number (n = 60) were serving in JDAL billets. If we look at the major billet

**Figure 4.1
Distribution of 0-4–0-6 Reserve Active-Status List Officer Incumbents by
Reserve Component**

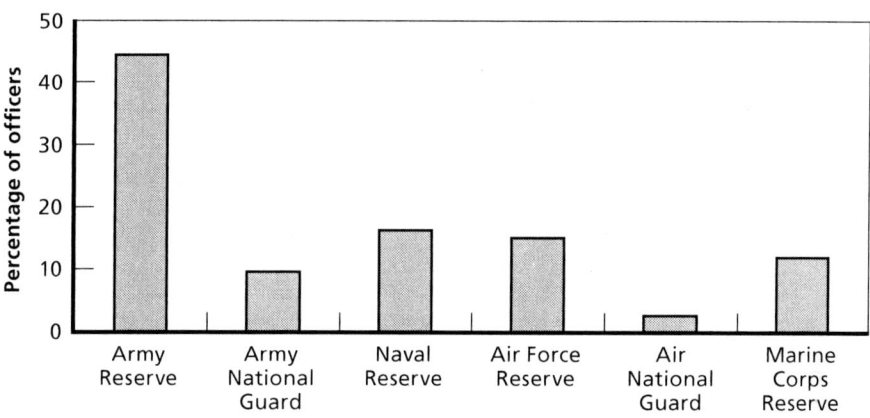

organizations, we find that 52 percent of reserve officers were serving in
two kinds of organizations: various CENTCOM JTF and geographic
commands, while another 9 percent are assigned to the U.S. Joint
Forces Command and 8 percent to the Joint Staff. Of the 99 officers in
service-nominated billets, between 32 and 36 are serving in Army and
Air Force billets, 24 in Navy billets, and 7 in Marine Corps billets.

Measuring the Jointness of the Billets

The survey gathered information on a number of characteristics of the
billet that are generally regarded as defining jointness. These include
types of tasks performed during a typical work week, supervision of the
billet by non-own–service or civilian personnel, frequency and number
of interactions with non-own–service organizations and personnel,
need for joint professional education or prior joint experience for successful job performance, and types of joint experience provided by the
billet. Obviously there are other measures of jointness, but these characteristics are a reasonable subset. We use officers' responses to these
questions to provide a broad-brush picture of how billets in which these
reserve officers are serving rank along these dimensions, as a means of
determining the need for joint officers to serve in these assignments.

Table 4.1
Selected Characteristics of Billets Whose Incumbents Are Reserve Active-Status List Officers, 0-4–0-6

Selected Characteristics	Percentage (n=510)
JDAL Status	
Billets currently on JDAL	11.8
Billets in external organizations with billets on the JDAL	
Billets nominated by the services	68.8
	19.4
Major Billet Organization Category	
Army	6.3
Navy	4.7
Air Force	7.1
Marine Corps	1.4
Joint Staff	8.2
Office of the Secretary of Defense (OSD)	--
CENTCOM JTF	28.4
International Organizations[a]	0.2
Combat Support Agencies (CSA)[b]	3.5
Other Non-OSD Defense Agencies[c]	1.0
OSD Defense Agencies[d]	--
Educational Agencies[e]	2.4
Geographic Commands[f]	23.9
Force Provider (*U.S. Joint Forces Command*)	9.2
Functional Commands[g]	3.7

[a] Inter-American Defense Board; North Atlantic Treaty Organization.

[b] Defense Contract Management Agency (CSA); Defense Information Systems Agency (CSA); Defense Intelligence Agency (CSA); Defense Logistics Agency (CSA); Defense Threat Reduction Agency (CSA).

[c] Joint Theater Air and Missile Defense Organization; Missile Defense Agency; National Geospatial-Intelligence Agency; National Security Agency; North American Aerospace Defense Command.

[d] American Forces Information Service; DoD Counterintelligence Field Activity; DoD Human Resources Activity; DoD Inspector General; Office of Economic Adjustment; Pentagon Force Protection Agency; TRICARE Management Activity; Washington Headquarters Services.

[e] Defense Acquisition University; National Defense University.

[f] U.S. Central Command; U.S. European Command; U.S. Northern Command; U.S. Pacific Command; U.S. Southern Command; U.S. Special Operations Command.

[g] U.S. Strategic Command; U.S. Transportation Command.

Table 4.2 defines the set of indicators used to characterize jointness.

Tasks Performed During a Typical Work Week

Four tasks were selected as representing "highly joint" activities: (a) providing strategic direction and integration; (b) developing/assessing joint

Table 4.2
Definitions of Indicators Used to Characterize "Jointness"

Metric	Indicator
I. Tasks performed during the typical work week	Percentage of officers providing strategic direction and integration Percentage of officers developing/assessing joint policies Percentage of officers developing/assessing joint doctrine Percentage of officers fostering multinational, interagency, or regional relations Percentage of officers performing three or more of these tasks
II. Interactions with non-own–service organizations and personnel	Median number of non-own–service organizations the officer interacts with monthly or more frequently Median number of non-own–service personnel with whom the officer interacts monthly or more frequently
III. Supervision of billet by non-own–service personnel/civilians	Percentage of officers reporting being supervised by one or more non-own–service supervisor/civilian/non-U.S. military personnel or civilian
IV. Need for joint professional education or prior joint experience	Percentage of officers reporting that JPME II is required or desired for the assignment Percentage of officers reporting that prior joint experience is required or desired for the assignment
V. Types of joint experience provided by the billet	Percentage of officers reporting getting significant experience in multiservice matters Percentage of officers reporting getting significant experience in multinational matters Percentage of officers reporting getting significant experience in interagency matters Percentage of officers reporting getting significant experience in all three areas

policies; (c) developing/assessing joint doctrine; and (d) fostering multinational, interagency, or regional relations. As is true with the larger sample of officers, reserve officers were much more likely to report doing the first task than the others, as shown in Figure 4.2. Over half of the reserve officers were providing strategic direction and integration, one-third were engaged in developing or assessing joint policies, and about one-fourth were doing each of the other two tasks. Overall, about 72 percent reported performing one or more of these tasks. Although sample sizes are small, we examined the percentage of officers in various billet organizations who reported doing one or more of these tasks. We combined all service-nominated billets and excluded organizations where fewer than 20 reserve officers were serving. This left us with billets in

Figure 4.2
Percentage of 0-4–0-6 Reserve Active-Status List Officers Performing Selected Tasks Characterized as "Highly Joint"

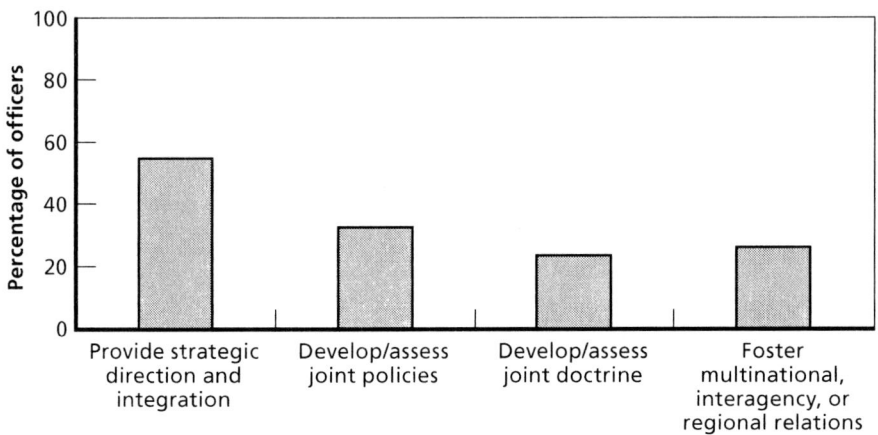

RAND MG517-4.2

five organizations: Service-nominated billets (n = 99); Joint Staff (n = 45); U.S. Central Command (CENTCOM) JTF (n = 144); Geographic Commands (n = 122); and Force Provider (n = 47).

Figure 4.3 displays the percentage of reserve officers reporting performing none or one or more of the joint tasks by these five billet organizations. Officers in internal service billets or at CENTCOM JTF were less likely to be engaged in these activities; more than 40 percent of those on the JS reported doing three or more of these tasks during a typical work week.

Frequency and Number of Interactions with Non-Own–Service Organizations and Personnel

As with the larger sample, we find that officers differ in the number of non-own–service organizations and personnel with whom they interact monthly or more frequently, depending on where they are assigned, as shown in Table 4.3. Overall, reserve component officers reported interacting frequently with four non-own–service organizations and

Figure 4.3
Percentage of 0-4–0-6 Reserve Active-Status List Officers Performing None, One or More Tasks Characterized as "Highly Joint," by Major Billet Organization

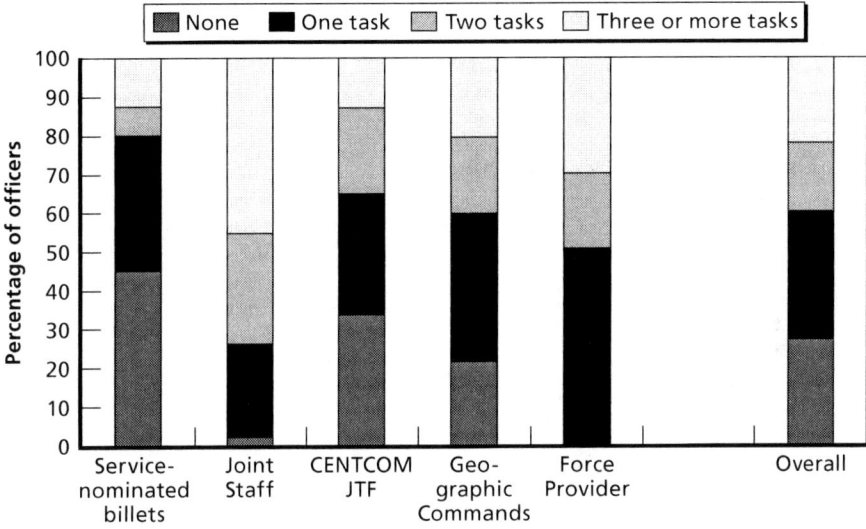

RAND MG517-4.3

Table 4.3
Median Number of Non-Own–Service Organizations and Personnel with Whom 0-4–0-6 Reserve Active-Status List Officers Interact Frequently, by Major Billet Organization

Selected major billet organization	Median number of non-own–service organizations officer interacts with monthly or more frequently	Median number of non-own–service personnel with whom the officer interacts monthly or more frequently
Service-nominated billets	2	3
Joint Staff	10	5
CENTCOM JTF	4	7
Geographic Commands	4	5
Force Provider	5	5
Overall	4	5

five non-own–service types of personnel.[6] Those assigned to internal service billets tended to have the lowest number of frequent interactions with non-own organizations or personnel (2 and 3, respectively). Those on the JS reported interacting frequently with 10 organizations; those in CENTCOM JTF billets reported interacting with seven different types of personnel. The findings are similar to those in the larger sample of active duty officers.

Supervision of Billet by Non-Own–Service Personnel

About 62 percent of reserve component officers overall reported being supervised by at least one non-own–service supervisor, but those in service-nominated billets were the least likely to do so (Figure 4.4). Less than a quarter of the reserve officers were being supervised by someone other than their own service superior, and this was similar to what we found in the larger sample. More than 80 percent of officers serving in force provider billets (U.S. Joint Forces Command) were being supervised by non-own–service personnel.

Need for Joint Professional Education and Prior Joint Experience for Billet Assignment

About 57 percent of reserve officers reported not having credit for JPME I; 80 percent had no experience with JPME II (not surprising, given that reserve officers are not afforded opportunities to take JPME II) (Figure 4.5). Overall, about 27 percent of officers had received credit for JPME I, and a small percentage—7 percent—had received credit for JPME II. As with the larger sample, a not insubstantial group of

---

[6] The survey participants were able to choose from among the following types of personnel: U.S. Army personnel, U.S. Navy personnel, U.S. Air Force personnel, U.S. Marine Corps personnel, U.S. Coast Guard personnel, other DoD civilian, other U.S. civilian, non-U.S. civilian, and non-U.S. military officer. In the case of the U.S. uniformed services, the option included all personnel military and civilian. For example, the U.S. Army personnel option was specified to include officer, enlisted, or civilian; Active-duty, National Guard, or Reserve.

Figure 4.4
Percentage of 0-4–0-6 Reserve Active-Status List Officers Being Supervised by Non-Own–Service Supervisors, by Major Billet Organization

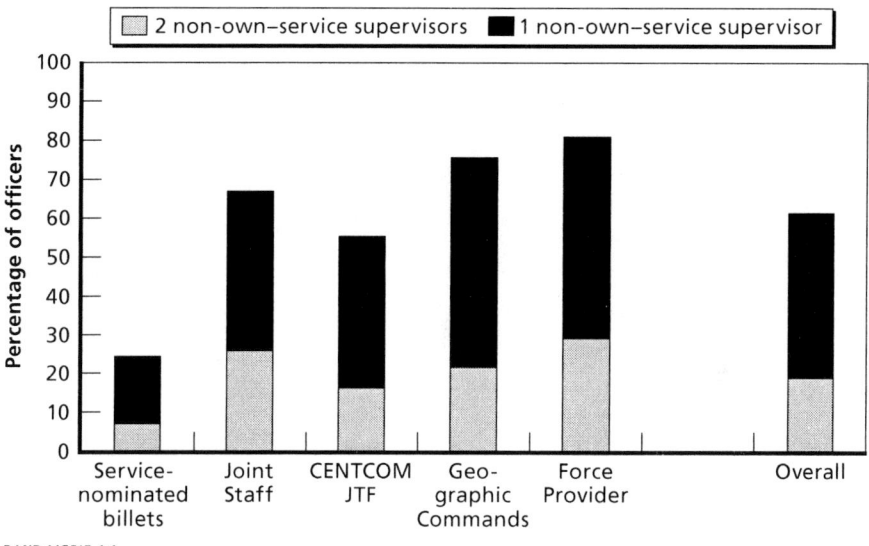

RAND *MG517-4.4*

Figure 4.5
Percentage of 0-4–0-6 Reserve Active-Status List Officers with Credit for JPME I and JPME II

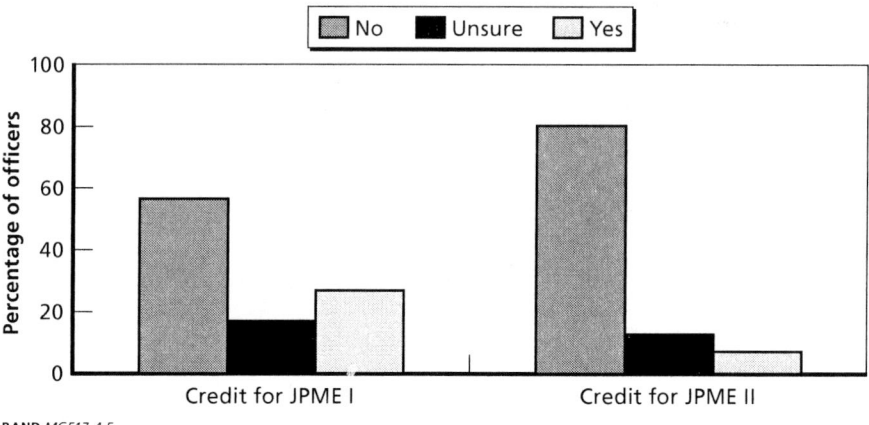

RAND *MG517-4.5*

Figure 4.6
Percentage of 0-4–0-6 Reserve Active-Status List Officers Reporting That JPME I, JPME II, Prior Joint Experience, or Other Joint Training/Education Was Required or Desired for Billet Assignment

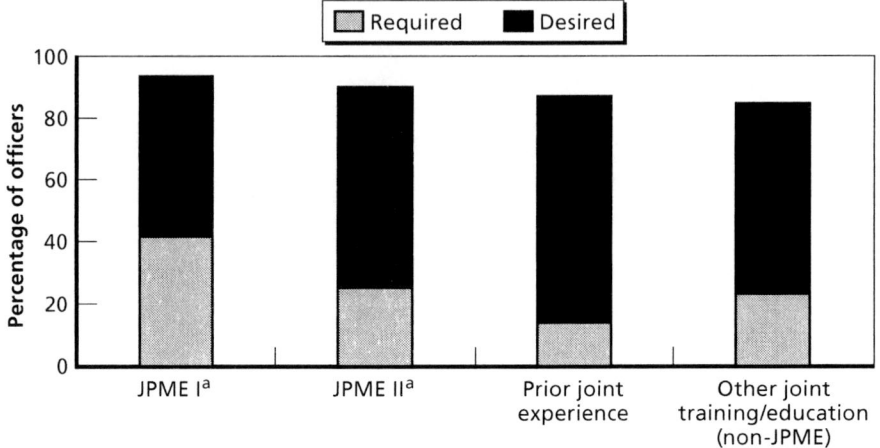

NOTE: [a]Of those with experience of relevant phase of JPME.
RAND MG517-4.6

officers seemed unsure of whether they had received credit for these two phases of JPME.

Officers were asked their opinion as to whether JPME I, JPME II, prior joint experience, and other joint training/education (other than JPME) would be helpful for successful performance in their assignment. Figure 4.6 shows their responses. When asked about joint professional education, joint experience, and joint training, the majority of officers believed that joint professional education, prior experience in a joint environment, and training were required or desired in order to perform their duties successfully.[7]

Officers reported that about 90 percent of the billets needed (required or desired) JPME II for successful performance (if we include only responses where officers had JPME II experience), and about 87

---

[7] We restricted the responses to those with experience with JPME when calculating the percentages who reported that JPME I or JPME II was required or desired. If all responses are included, then 46 percent of billets need JPME II.

percent of the billets needed prior joint experience. A small subset of billets, about 5 percent (not shown in Figure 4.6), were coded as requiring both JPME II and prior joint experience, thus defining billets that would need to be staffed by fully joint qualified officers. Also, officers in internal service billets were less likely to cite need but even among them, between 68 and 80 percent believed such education and experience to be required or desired for job performance. Among other major billet organizations, well over 80 percent of officers reported that joint education and/or experience were required or desired for the assignment.

Implications for Estimating Demand for Reserve Component Joint Officers

It should be amply evident from the discussion above that the majority of billets in which reserve component officers are serving would benefit from joint education and/or prior joint experience, and that some appear to require such qualifications—the crux of joint duty assignments.[8] As we mentioned, we suspect that these billets were staffed by IMAs and AGRs. Thus at a conservative estimate, about 5 percent of all billets staffed by IMAs and AGRs will require fully joint qualified officers, whereas between 46 and 87 percent will require JPME II or prior joint experience. However, we had also shown that billets in external organizations were much more likely to require joint education or prior joint experience; thus it is important to validate the need for officers with advanced JPME or prior joint experience through a billet-by-billet analysis. We have little information on the need for joint education or experience for the majority of unit reservists, although one could conjecture that there is likely to be little demand for joint education or experience among the vast majority of them. However, this runs counter to the position adopted by the National Guard, which states that all guardsmen should be trained in joint matters, and to the Chairman of the Joint Chiefs of Staff (CJCS) Vision for Joint Officer Development that all O-6 should be Joint Qualified Officers. This may be near a reality for many AGRs and IMAs but not for unit reservists.

---

[8] Many of these billets also provide joint experience, as discussed in the next chapter.

In round numbers, in the grades of O-4 and above there are around 10,000 AGRs (about half are in drilling units), 10,000 IMAs, and 50,000 drilling unit members. Emphasis should be placed on determining the need for prior joint experience and education among the 5,000 nondrilling unit AGRs and the IMAs. Organizations that have these billets (especially OSD, JS, and Combatant Commands) should be tasked, as a one-time data collection, to specify the prior needs for joint education and experience. Job specifications (requisitions) may already contain such a notation. Moreover, a sample survey of position incumbents could buttress the organizational assessment and gain further information about experiences in representative positions from a far greater range of positions. For example, how does demand differ by grade, occupation, reserve category, duty status, reserve component, and other characteristics of interest. Eventually databases should contain demand information. Either way, however, understanding how many reserve active-status list officers need to be developed as joint officers or joint qualified officers requires a determination of demand for them.

Summary

Neither the JDAL for active duty list officers nor the JDA-R for reserve active-status list officers shows where joint experience (or education) might be needed. Although some services and components may have begun to make these determinations, we found no databases that routinely collected it. Both the JDAL and the JDA-R are designed to list positions that qualify officers with joint experience. The supply of officers with joint experience is discussed in the next chapter.

CHAPTER FIVE

Determining Potential Supply of Reserve Active-Status List Joint Officers

This chapter pertains to the supply, or provision, of officers qualified with jointness. The current joint officer management system, and the related discussion and debate, generally focuses on credit for officers based on their completing certain activities rather than qualifications of officers themselves. As discussed in Chapter Two, it is a process evaluation rather than an outcome evaluation. An ideal system would focus instead on the qualifications of officers in the system for the good of the system. Such a system would acknowledge those officers who are joint qualified and thus grant them "credit," but the focus would be on qualifying them as needed by military organizations, or by the military overall. Thus the intent of this section is to explore different ways that officers could be qualified for future joint assignments or for other military purposes (to include the development of joint qualifications for generals and flag officers).[1]

There are four different processes that govern officer qualification: joint assignments, joint education, joint training, and joint acculturation. Most discussion of joint qualification focuses on the first three of these: assigning, educating, and training. This report focuses primarily on joint assignments and joint education. Training typically provides

---

[1] Indeed, the demand for joint qualified officers, or the extent to which qualified officers are needed, could vary, based on (a) the demand for joint officers in specific jobs, (b) the extent to which the officer system should emphasize joint and thus have field grade officers throughout the system with prior joint experience and joint perspectives or joint knowledge, and (c) the likelihood that certain officers would be future candidates for general or flag officer and thus require joint experience before promotion to those pay grades.

job-specific capabilities rather than the broader, more general expertise or knowledge of education.[2] Thus training officers for joint assignments is similar to training officers for service-specific assignments; we will assume that officers are properly prepared with job-specific capabilities regardless of the organization to which they are assigned. We will also assume limited benefit to later assignments or subsequent organizations unless similar to previous ones. Acculturation to the joint environment is rarely mentioned explicitly in joint officer management, but it is the underlying basis for some concerns and constraints. For example, officers were, until recently, only able to obtain the second phase of Joint Professional Military Education (JPME II) from educational organizations that were managed jointly. This is, in part, because of the acculturation that officers are expected to obtain during their education. Similarly, the current restriction that active duty list officers may obtain joint experience only from positions external to their own service is, in large part, an assertion that officers must acculturate to the joint environment before receiving joint credit.[3]

This chapter focuses on joint assignments and joint education as the primary processes that qualify a reserve active-status list joint officer. Thus the question of how officers might receive joint credit actually becomes a question of how officers might gain valid joint experience (and thus an administrative notation that they have joint credit) and joint knowledge. However, credit for valid joint experience is just a single aspect of how to develop and accredit, and thus how to make available to organizations, joint qualified officers. (As stated in Chapter Two, we use the term *joint qualified officer* as one having experience, knowledge, and acculturation in joint matters. This term is inclusive of but broader than the term *fully joint qualified* as specifically defined in Department of Defense Instruction (DoDI) 1215.20. In other words,

---

[2] To the extent that training involves frequent association with officers of other services, agencies, and nations, a certain amount of acculturation does take place that could be included in assessing overall qualifications.

[3] Officers who associate with officers of other services and nations and with officials of other agencies gain an understanding of underlying values, attitudes, and beliefs. Such acculturation should improve the "fit" of officers with joint (multiservice, multinational, and interagency) organizations.

an officer might be fully joint qualified and thus a joint qualified officer by meeting the requirements in the instruction (process measurement), but there may be other means (outcome measurement) for also becoming a joint qualified officer. These other means are discussed below. Thus, the following discussion explores how reserve active-status list officers might be assigned and educated in joint matters or have their knowledge and education measured and validated, so as to provide joint qualified officers (with requisite joint knowledge, experience, and acculturation) for organizations that provide needed military capabilities.

As discussed in Chapter Two, we recognize that management of reserve active-status list officers is more of a "pull" system than a "push" system. There are not central assignment processes as there are for active duty list officers. As a result, individual officers must be heavily involved in decisions about their assignments and education. In the words of an Air Force officer, the system "develops those who want to be developed." The concept of an opt-in system of reserve joint officer management is to be explored. Particularly, this might be useful for those who aspire to general or flag officer. Moreover, the assigning and educating processes, to the extent they exist, vary for the different reserve categories: unit-based officers, individual mobilization augmentees, and active guard and reserve.

The current joint officer system for active duty list officers is focused almost exclusively on joint credit, which should ideally be just the administrative result of officers receiving valid joint experience. Debate and questions about how to obtain joint credit for both active duty list and reserve active-status list officers can be reduced to a debate of which circumstances, whether formal assignment or ad hoc responsibilities, provide a valid joint experience for officers. The active component system is currently constrained by the Goldwater-Nichols Act (as contained in Title 10 US Code) as far as the types of assignments (and the tenure required in those assignments) before an officer receives joint credit.

Reserve component joint officer management is not so constrained. Whereas some advocates argue that only the adoption of a similar (if not identical) system to provide reserve active-status list officers with joint duty credit will provide them with the credibility necessary to serve with active duty list officers, others argue that the active JOM is inadequate (if not entirely dysfunctional) and that the reserve component has the opportunity to adopt a system considerably superior to the active JOM. The intent of this report is to explore the possibilities available to a reserve component JOM and conceptualize how they might be implemented.

Systems for Determining When Officers Are Joint Qualified—What Is the Supply of Joint Officers?

This section discusses different systems that could be used for evaluating either processes or outputs/outcomes as a basis for qualifying officers. We first discuss options for using a time- and billet-based system and decisions to be made about the processes for doing so. We then discuss methods for measuring experience of individual officers from an output/outcome orientation.

A Time- and Billet-Based System for Determining When Officers Obtain a Joint Experience

There are many options to consider, and the resulting system need not be constrained to a single option. For example, a time- and billet-based system is the most administratively simple alternative because it specifies measurement rules. Such a system could be similar to the active duty Joint Duty Assignment List (JDAL), such as the prescribed joint duty assignment-reserve (JDA-R) list. Other simple rule-based process alternatives also exist. For example, should a reservist serve in an active duty JDAL billet, the reservist could receive joint credit as though he had served in a JDA-R billet. (This is currently precluded by law.) Or, if reservists serve in temporary billets (e.g., in a Joint Task Force [JTF]) side-by-side with active duty list officers who receive joint credit, a "running mate" concept could be used to award credit to the reservist. When billets are identified as providing a valid joint experience, then officers who serve a minimum-length assignment are assumed to have gained a valid joint experience, whether active or reserve.

Such a billet-based system requires some decisions, however, some of which are listed below and subsequently discussed:

- Should included billets be only in organizations external to their own service?
- Should included billets be only in organizations external to their own reserve component?
- Should all billets in selected organizations be included?
- Should all officers be required to spend the same tenure to obtain joint credit?

Unlike for active duty list officers, neither law nor DoD policy currently precludes credit for in-service billets, given the billets meet prerequisites. DoDI 1215.20 allows for assignments to organizational positions other than Office of the Secretary of Defense (OSD), the Joint Staff (JS), and Combatant Commands to be on the JDA-R if the incumbents are involved with the integrated employment or support of land, sea, and air forces of at least two military services. Moreover, the incumbent's duties shall deal directly with duties relating to the national military strategy (NMS), joint doctrine or policy, strategic planning, contingency planning, or command and control of operations in support of a Combatant Command. There is no prohibition against in-service billets being included on a JDA-R, given the billets meet the stated requirements.

Is there a downside to including in-service billets on a JDA-R? "Two services" is a lesser test than the "two military departments" test of the active component DoDI. Thus it appears that a Navy reserve active-status list officer in a position involved with Navy and Marine Corps employment or an Air National Guard (ANG) officer in a position involved with Army National Guard (ARNG) and ANG employment would qualify. If internal service organizations become a source of joint credit, then some officers (e.g., future leaders) might not seek assignments in external organizations. Also, there might be a question of how much association with officers of different services (which leads to acculturation) might occur in such positions. There are likely internal service organizations with billets that, either due to their daily responsibilities or due to the associated temporary assigned duty/temporary duty (TAD/TDY) responsibilities that emerge during military missions (e.g., JTF duty), provide assigned officers with a valid joint experience. If such billets meet the stated requirements, they should be considered for inclusion on the JDA-R. One could argue that the joint experience in such positions is less valid than in external organizational positions, but there is a tradeoff between such an argument and the low administrative cost of adding service billets to an existing JDA-R system.

If in-service billets are included on a JDA-R, it is likely that an individual mobilization augmentee (IMA) within a service component of a combatant command could qualify if the duties meet the test. More problematic is a reserve unit, part of a service component command, that supports a combatant command. In some cases, the members of the unit might be used as individuals within the service component or combatant command headquarters or in a JTF; in other cases the entire unit might be employed in support. If the latter case, association of most officers in the unit might be with officers of their own service reserve component. Another problematic example would be an Army reserve unit in support of an Army National Guard operation as part of a U.S. Northern Command (NORTHCOM) JTF. Although lots of interagency experience might be gained in such a situation, the current instruction does not appear to allow for this type of within service and within component or within service but across component experience. The nuances of reserve component category, reserve component duty status, and reserve component employment introduce a level of complexity in crafting administrative rules for including positions in "other organizations" on a JDA-R. The answer is that other means to measure and validate are needed when simple administrative rules cannot cover all situations. This is explored later in the chapter.

Whether all reserve component billets in selected organizations should be included on the JDA-R has both administrative implications and broader officer management implications. The most simple management solution is to select organizations for inclusion on the JDA-R, and then include all billets in those organizations. This reduces the administrative burden of assessing individual billets, as they change over time, for continued inclusion or exclusion from the JDA-R. The DoDI singles out billets in OSD, the JS, and the Combatant Commands, and arguably all positions in these organizations would meet the associated "duties" test specified in the instruction. Managing with such a minimum administrative burden, however, will likely provide individuals with joint credit who have not received a valid joint experience, and thus will not provide an accurate portrayal of which officers have valid joint experiences. Nonetheless, it appears that administrative simplicity is the better course for these positions. Providing credit to all positions for reserve active-status list officers in OSD, JS, and Combatant Commands greatly benefits AGRs and IMAs, as they are the two categories of reservists who serve in these positions permanently. Far fewer, if any, unit reservists would gain credit this way.

Must all officers remain in a billet for the same tenure to receive joint credit? Administratively, the simplest answer is yes. A system that requires officers to stay in their position for a predetermined amount of time is the simplest, and thus the least costly, to manage; however, theory and current experience suggest otherwise. There exists within DoD some current acknowledgment that "intensity" may be a factor to consider in conjunction with tenure. The joint officer survey collected proxy data for intensity, asking questions such as the number of weeks the officer was working; how quickly the officer acclimated to joint duties;[4] whether the officer was serving at his home base; or receiving family separation allowance, special pay for duty subject to hostile fire or imminent danger, special pay for hardship, and so forth. The intensity argument asserts that officers performing under extreme conditions are more likely to retain that experience. Although the effect of stress is unclear on the retention of learned experience, one can argue that an officer working 12-hour days is likely gaining his joint experience

---

[4] On average, officers in JDAL billets reported that it took about 5 months to become comfortable operating in a joint environment. It was a little shorter for those in non-JDAL billets in external organizations—about 3 months. The 25th percentile was 2 months and the 75th percentile was 6 months, so the middle 50 percent of officers reported that it took between 2 and 6 months to become comfortable in a joint environment. Those assigned to CENTCOM JTF billets reported becoming comfortable working in the joint environment within a short period of time—1 month.

more quickly than an officer working 8-hour days. Further, the strongest correlation between experience and job performance—in this case whether a joint experience bears on future job performance—occurs when the measurement is the amount of times an individual has performed a task.[5] Thus longer work days in more intense circumstances suggest that a valid joint experience is gained with shorter tenure than in a routine office environment. Although 2 years of full-time support service and 3 years of non-full–time support service would qualify officers in designated positions (which can be waivered to 18 and 30 months, respectively), there should be provisions for officers to qualify with less time given more intense or frequent duties than normally expected.

What about the usual tenure of reservists other than AGRs and IMAs? Should a reservist obtain joint credit for military service, even that in a joint environment, that is measured in weeks rather than months? The DoDI allows for cumulative credit. Such joint experience should be noted in a reservist's file, but the existing data systems are not likely to track such information. An organization (e.g., a Standing JTF Headquarters) might likely be interested in obtaining the assignment of a reservist who has experienced multiple short-tenure joint assignments, as compared to a reservist who has not been exposed to a joint environment at all. It is unlikely, however, that individuals with multiple short-term joint experiences would be considered to have obtained sufficient joint experience (3 years cumulative) to be fully joint experienced. This suggests the need for different levels of joint accreditation to acknowledge all officers who have been exposed to a joint working environment without accumulating sufficient experience to become fully joint experienced. Such a system would greatly assist unit-based officers, who might have short periods of active duty for special work with joint organizations, to gain credit. These different levels are discussed in more detail later in this chapter.

In summary, a time- and billet-based system has both strengths and weaknesses. The strengths are administrative; a billet-based system

---

[5] See the discussion of the relationship between experience and job performance in Appendix B.

can be the easiest system to manage, requiring relatively simple updates to maintain a mostly accurate system. Such a system also manages expectations of individuals and eases the jobs of those responsible for assessing qualifications of future leaders; if officers are assigned to specific jobs for requisite time, then they can be assured that they will receive joint credit at the conclusion of an assignment. However, a time- and billet-based system has weaknesses as well. When the system is designed to ease the management burden, there are officers who gain a joint experience but do not receive credit because their billets are not on the list. Alternatively, there are officers who receive joint credit without having an appropriate joint experience. Such a system is inherently less adaptable to changing circumstances, or temporary or evolving joint missions. However, a time- and billet-based system can be the primary basis of a reserve active-status list joint officer management system and could be supplemented by other approaches.

Individual Experience-Based Methods for Determining When Officers Obtain Joint Qualifications

Theory and research provide many different ways that rating methods can be used to determine when individuals have obtained a joint experience. Typically, these methods determine not just whether an individual has valid experience, but also the extent of the individual's training and education. Thus, individual-based methods might be used in conjunction with a billet-based system in at least three ways:

- The first case is reflected in the current JDA-R approach, where a time- and billet-based system determines when officers have obtained JDA-R experience and an individual-based minimum qualifications system determines when officers should be fully joint qualified. In such instances, the individual-based system might use the result of the time- and billet-based system as input, to establish valid experience for individuals, regardless of which individual-based system was used.
- Second, a revised system might not depend exclusively on the time- and billet-based system to recognize valid joint experience.

- A third approach might depend entirely on an individual-based system and disable the billet-based JDAL system.

Before evaluating how best to combine (or not) individual-based systems with a billet-based system, it is worthwhile to note the relative advantages or disadvantages of different individual-based methods. As Appendix B describes in more detail, there are many methods available for individual officer evaluation of knowledge and experience, including minimum qualifications methods, point methods, employment and assessment tests, self-evaluating task methods, and methods that rely on accomplishment records. These methods vary in the administrative effort required, whether the individual portrays his own qualifications, generalizability across individuals, and validity in the ability to predict subsequent job performance. The current JDA-R system outlines a minimum qualifications method to certify individuals as fully joint qualified. In other words, all officers who have accumulated the minimum required amount of joint experience (in the billet-based system) and have completed the minimum required joint education are certified as fully joint qualified. However, research indicates that minimum qualifications methods are not generalizable; it is difficult to equate education and experience across individuals. Further, the validity of minimum qualifications methods is not known.

Point methods might identify jointness among officers by assigning officers "points" based on the length or other aspects (such as intensity, location, etc.) of their joint assignment and the joint education and training they had received. Point methods are widely used and provide a relatively easy and inexpensive method to assess individuals. Point methods could be improved by considering different levels of qualification for officers; given revised criteria for qualification, points could identify minimally qualified joint officers as well as those who were fully joint qualified. However, point methods generally exhibit a low mean validity and a lack of generalizability as well as low inter-rater reliability when qualifying in groups.

Tests and formal assessments can validly evaluate an individual's professional expertise and thus acknowledge experience and knowledge gained. They can also predict future performance with higher validity than most other methods. However, those methods are only as good as the test or assessment vehicles that they depend on. Creating, evaluating, and maintaining these tests or assessments can be very expensive. Given changing world events and many different occupations, developing and maintaining current testing vehicles may be extremely expensive, and administering these tests or evaluations could be very time-consuming.

Task methods depend on self-evaluation; individuals both identify the amount of time they typically spend on identified tasks and evaluate their proficiency in tasks. Although there are recommended ways to reduce the risk of purposely inflated answers and to increase the validity of this method, task methods are still problematic.

Individual evaluation based on accomplishment records provides the most valid method to assess individual qualifications and is less susceptible to answer inflation or individual faking. This method would require individuals to document major accomplishments that demonstrate their proficiency in identified joint areas. The areas of required proficiency would need to be determined, and the criteria for determining acceptable proficiency would need to be determined. The method and process of evaluating and approving these petitions for joint credit would need to be determined. For example, evaluation of these petitions for joint credit could occur locally and might be endorsed or denied by a two- or three-star general or flag officer, to be forwarded to the JS or to a board created and managed by the JS. Evaluation might also occur centrally by a board or committee representing the JS. Regardless of the evaluation process, this method could validly acknowledge individuals who have received a joint experience in a nontraditional billet or circumstance.

It appears that either a point method or the accomplishment record is most tractable as a means to evaluate knowledge and experience of individual officers. Both have antecedents in personnel management practices for reserve active-status list officers. Both reduce the administrative onus on the system by making the officer responsible for initiating and documenting knowledge and experience if they are determined to be qualified.

A Combined Billet- and Individual-Based System to Determine a Valid Joint Experience

The methods above provide the building blocks for a more flexible, accommodating, and valid joint officer management system. If a billet-based and tenure-limited structure serves as the primary element for identifying joint experience, then the JDA-R could contain those assignments that are judged to consistently provide each officer who performs that job with a valid joint experience. Such a time- and billet-based system is administratively simple, albeit relatively inflexible. The required tenure might be consistent for all jobs, or it might vary by location or by characteristics of the job. However, there would be less pressure for the JDA-R to include all billets that might possibly, under some conditions, provide a joint experience, as the billet-based system could be complemented by an individual evaluation method that acknowledges the joint experience gained by officers in other, non–JDA-R assignments.

An individual-based system could provide joint credit to those officers who document their proficiency in identified joint areas. This system might consider the valid joint experiences of officers who were serving in civilian positions[6] or positions in organizations external to the service that were not on the JDA-R, or who served in JDA-R assignments for less than the identified required tenure. Such a system might also acknowledge the valid joint experiences of officers who were serving in service-specific organizations if such billets were not on the JDA-R. The evaluation criteria that assess the depth or breadth of experience required might vary for those officers serving in internal service positions, or the level of authority required to approve the joint experience of officers in internal service assignments might vary from those in external organizations, but this system could flexibly acknowledge officers who received a valid joint experience in assignments that were otherwise not typically or consistently joint. To the extent that officers

---

[6] DoD surveys reservists to determine their civilian skills.

in certain assignments were consistently applying for, and receiving, joint credit, such assignments should be considered for addition to the JDA-R.[7]

Different Levels and Different Types of Joint Experience or Proficiency

Regardless of the combination of accreditation structures used to identify officers who have received a valid joint experience, joint education, joint training, or even joint acculturation, there should be recognition in the system of different levels of joint proficiency. For example, while it is important to acknowledge those officers who are fully joint qualified, it is also important to recognize those officers who have received sufficient joint education to begin a joint position with some proficiency. It would also be useful to acknowledge those officers who have received some joint experience, even if they served for an insufficient tenure to receive full credit. These differing levels of qualification could include joint officers and joint qualified officers as introduced in Chapter Two. Joint officers are those who can demonstrate knowledge gained through education, self-study, some joint experience, or by other experiences. Fully joint qualified officers would have a rich combination of considerable knowledge and joint experience gained through some combination of JPME II, assignment of prescribed tenure in a billet-based system, or expertise evident and demonstrated through other qualification systems.

Such proficiency levels could also acknowledge the relatively different levels of experience gained by reserve active-status list officers. While some reservists serve as AGRs and can become fully joint qualified even with tenure requirements designed for active component officers, other reservists are more likely to serve consistent with traditional

---

[7] The process and outcome of providing joint credit to officers in an internal service organization should be carefully monitored and assessed, perhaps in a pilot study to determine what kinds of officers were receiving joint credit and for what kinds of job experiences. Moreover, the discussion, while focused on JDA-R and non–JDA-R positions, is applicable to JDAL and non-JDAL positions for active duty list officers.

reserve expectations. For such reserve officers who fill a joint assignment for several weeks, recognition that they are joint experienced as reservists would note their experience and make it easier to identify them for appropriate use, but would not confuse them with fully joint qualified officers.

Additionally, there may be value in separately identifying those officers who are proficient and experienced in multinational issues, multiservice issues, interagency issues, and intergovernmental issues.[8] As discussed in Chapter Three, the reserve component may have relatively more opportunities to participate in intergovernmental operations and thus both require and obtain such experience. They may also have considerable interagency opportunities but relatively few multinational opportunities, given their responsibility for domestic missions. Moreover, a system might want to consider the civilian occupations and duties of reserve active-status list officers. For example, an officer employed by a multinational corporation or by a federal, state, or local agency or government or by a contractor as a joint program manager might have considerable experience and knowledge to be valued.

Aligning Vision for Joint Officer Development with Active Component and Reserve Component Joint Officer Management Directives/Instructions[9]

In Chapter Two, we introduced the belief that there may be an unintended, but logical, underpinning that can be ascertained from these three publications that allows for a reconciliation of all three. Moreover, understanding how they might fit together helps us understand where change, if desired, could be made. Our interpretation is in Figure 5.1, and a discussion of it follows the figure.

To begin there is a set of service-qualified officers typically at the grades of O-4 to O-6. Active duty list officers can complete JPME II

---

[8] However, doing so should be decided after an assessment of the number of officers needed with these particular backgrounds. Small numbers may not warrant separate consideration.

[9] CJCS Vision for Joint Officer Development; DoDI 1300.20; DoDI 1215.20.

Figure 5.1
Rationalizing Vision and Directives

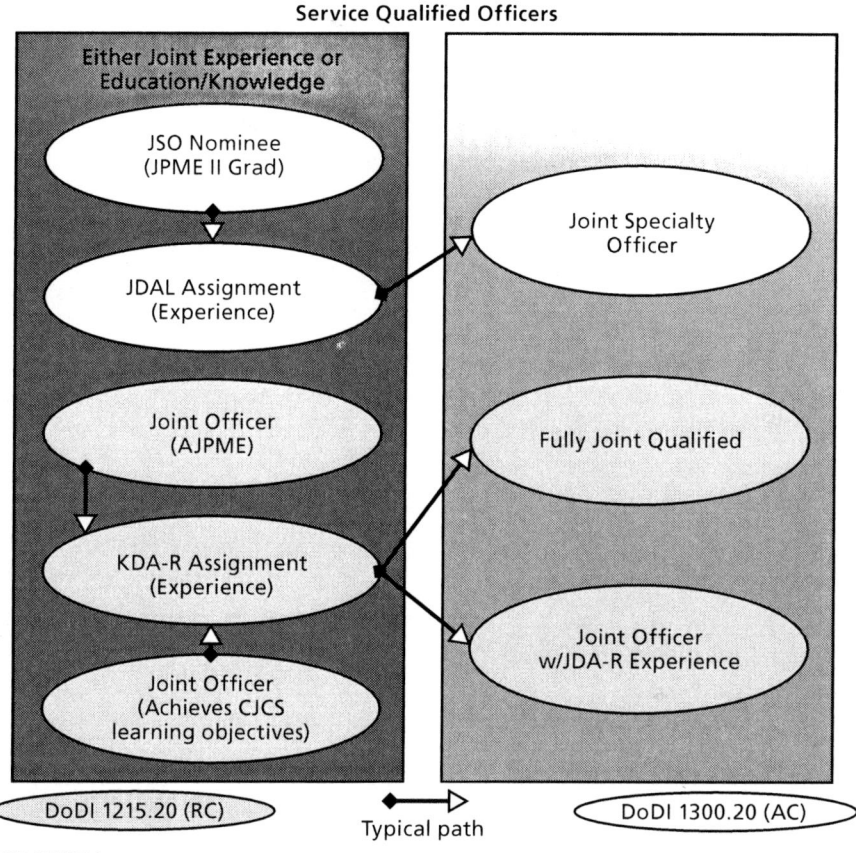

Service Qualified Officers

Either Joint Experience or Education/Knowledge

JSO Nominee (JPME II Grad)

Joint Specialty Officer

JDAL Assignment (Experience)

Joint Officer (AJPME)

Fully Joint Qualified

KDA-R Assignment (Experience)

Joint Officer w/JDA-R Experience

Joint Officer (Achieves CJCS learning objectives)

DoDI 1215.20 (RC)   Typical path   DoDI 1300.20 (AC)

RAND *MG517-5.1*

and become JSO nominees, or they can serve in a JDAL position for the needed time and receive credit for a joint assignment. Neither would be a Joint Qualified Officer by the standards of the Chairman of the Joint Chiefs of Staff (CJCS) Vision. Reserve active-status list officers could complete a JDA-R assignment or, without having such an assignment, complete Advanced Joint Professional Military Education (AJPME) or achieve CJCS learning objectives by other means. Neither would be a Joint Qualified Officer by the standards of the CJCS Vision. But these officers, active and reserve, have experience or knowledge qualifications that allow them to claim access to most joint assignments, to joint education, or to certain senior positions. An active duty list officer JSO nominee who completes a JDAL assignment or an officer who has completed a JDAL assignment who completes JPME II (under certain circumstances) or who completes a second JDAL assignment (under certain circumstances) would become a Joint Specialty Officer under DoDI 1300.20 and a Joint Qualified Officer by the standards of the CJCS Vision. A reserve active-status list officer with a JDAR-assignment who completes AJPME or otherwise achieves CJCS learning objectives becomes respectively a Fully Joint Qualified or a Joint Officer under DoDI 1215.20 and a Joint Qualified Officer by the standards of the CJCS Vision. Moreover, a Joint Officer without JDA-R experience can become both a Joint Officer and a Joint Qualified Officer by completing a JDA-R assignment.

Although this seems complicated, it is not so complex that it could not be tracked in databases and managed. It is a straightforward set of rules that is already in place with the CJCS Vision for joint officer development overlaid. However, although this set of rules seems sufficient to qualify many active and reserve officers, it does not appear complete with respect to reserve active-status list officers, especially those officers who face time and geography constraints for assignments and formal education. The CJCS Vision allows for achieving CJCS learning objectives without a formal school "diploma" but does not specify how this is to be done. This is an area for applying an output/outcome measure or evaluation to individual officers, as discussed previously. Moreover, if a point or accomplishment record approach is taken for knowledge, there is no reason why it could not be taken for measuring gained experience. We suggest that DoDI 1215.20 should be modified to allow for such procedures, and a pilot test of them could be undertaken. Not constrained by law, OSD could begin this process. Given the utility of this approach for reserve active-status list officers, such an approach might be extended to active duty list officers, but this would require legislative change.

Estimating the Current Supply, or Availability, of Reserve Joint Officers

Although the reserve joint officer management system has not been enacted to administratively identify reserve officers with joint experience or joint education, it is possible to estimate the number of reserve officers who have obtained a valid joint experience or who have attended advanced joint education through other data sources.[10] We can make such estimations, in part, from a survey of joint officers conducted in 2005 that focused on active component officers but incidentally included 679 reserve officers serving in assignments that were prejudged likely to either provide them with a joint experience or would have benefited from their having prior joint experience.[11]

Joint Experience Provided by Billet

In the earlier report (Kirby et al, 2006), we had seen that JDAL billets and non-JDAL billets in external organizations provided the most experience in multiservice, multinational, and interagency matters, and we see the same pattern here if we look only at reserve officers who were incumbents (Figure 5.2). Overall, more than 80 percent of billets in which reserve component officers were serving provided significant experience in multiservice matters, 63 percent provided significant experience in multinational matters, and 71 percent provided significant experience in interagency matters. Internal service billets were the least likely to provide significant experience in these three areas compared with other billets, but even among them, 50 percent provided significant experience in multiservice and interagency matters. Almost all Joint Staff billets provided significant experience in multiservice matters, and 90 percent of U.S. Central Command (CENTCOM) billets provided significant experience in multinational matters. More than three-fourths of billets in the geographic commands and CENTCOM

---

[10] It is not possible, however, to easily determine the officers who have obtained both joint experience and education, or who—in other words—are fully joint qualified.

[11] The RAND Joint Officer Management Survey is further documented in Kirby et al., 2006.

Figure 5.2
Percentage of 0-4–0-6 Reserve Active-Status List Officers Who "Agree/ Strongly Agree" That Their Assignments Provide Significant Experience in Multiservice, Multinational, and Interagency Matters, by Major Billet Organization

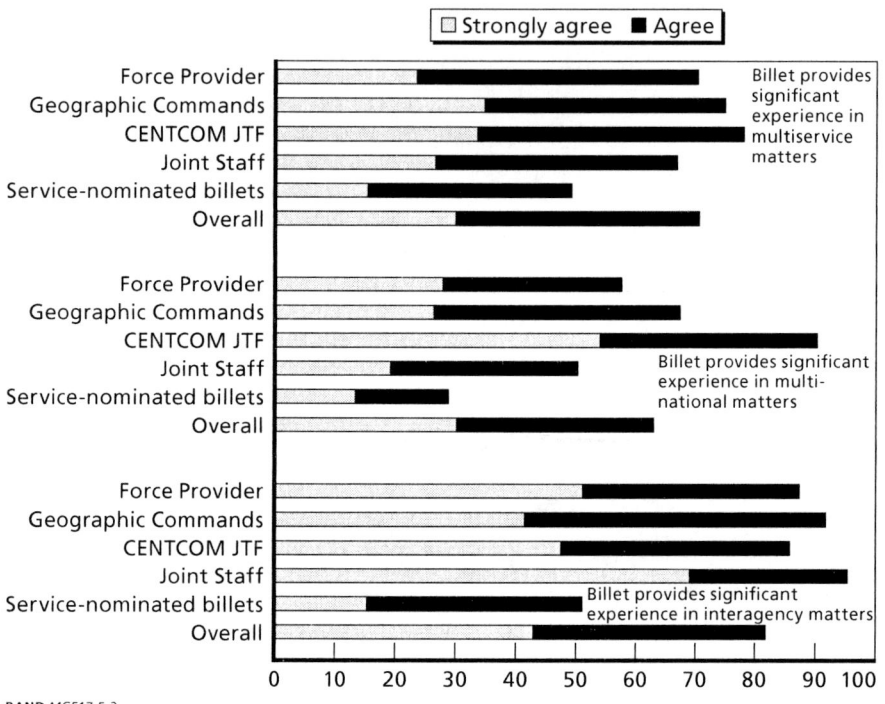

RAND MG517-5.2

Joint Task Force (JTF) involved experience with interagency matters, as did 70 percent of billets in the U.S. Joint Forces Command and two-thirds of Joint Staff billets. These responses suggest that the billets in which reserve component officers are serving mirror those in which active duty officers are serving in terms of the experience they provide their incumbents.

Figure 5.3 shows the percentage of billets that provide significant experience in one or more joint areas (multiservice, multinational, interagency matters). Overall, almost 90 percent of the billets to which reserve officers were assigned provided significant joint experience in at least one area and over half provided significant experience in all three

Figure 5.3
Percentage of 0-4–0-6 Reserve Component Officers Reporting That Billet Provides Significant Experience in One or More Areas—Multiservice, Multinational, and Interagency Matters

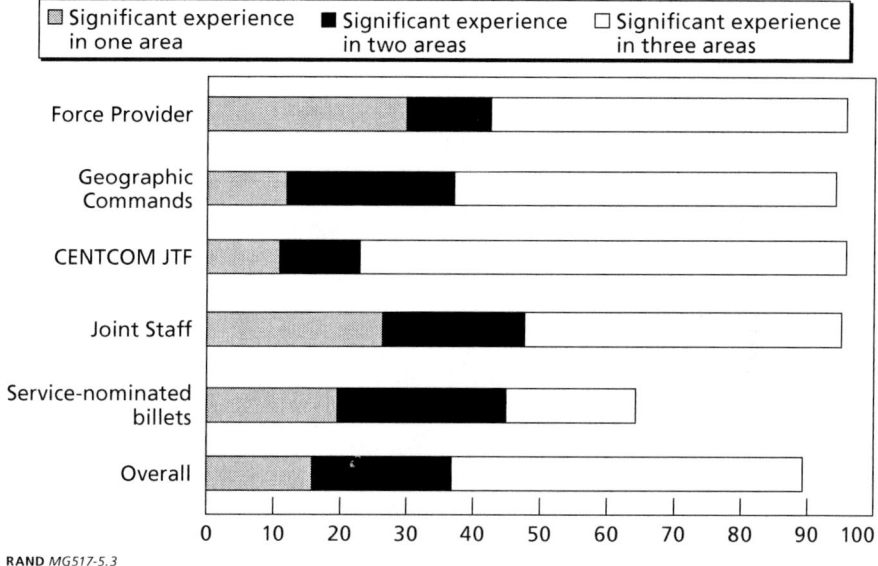

RAND MG517-5.3

joint areas. With the exception of the internal service billets, about half or more of billets in other organizations provided significant experience in all three areas. Billets in CENTCOM JTF were particularly likely to do so—close to three-fourths provided their incumbents with experience in all three joint areas.

Summary

Just from this data set we know that there are more than 500 reservists currently serving in billets who were surveyed. Of these, approximately 410 reserve officers believe they are gaining multiservice experience, 315 officers believe they are gaining multinational experience, and about 350 officers believe they are gaining interagency experience.[12] Of the officers surveyed, relatively few of them had received first phase of Joint Professional Military Education (JPME I) (approximately 20 percent), and even fewer had received JPME II (approximately 5 percent). However, given that there exist reserve officers who have completed or partially completed joint education, they should be tracked to ensure that they are best assigned to capitalize on that education. We know, for example, that approximately 225 reserve officers have completed AJPME through 2005, that approximately 250 and 400 reservists will complete AJPME in 2006 and 2007, respectively, and that JFSC plans to provide AJPME for as many as 500 reserve active-status list officers on an annual basis. We cannot, however, estimate with current data the number of reserve officers who have obtained a joint experience or joint education and who remain reserve officers. Nonetheless, given that the estimations of demand for joint-qualified reserve officers are as high as those discussed in Chapter Four, it is likely that there is currently an insufficient amount of reserve active-status list officers with joint experience and education, even if they are tracked and assigned appropriately to maximize the utilization of reservists who might be considered joint experienced, joint officers, or fully joint qualified, were such assessment practices in place. The relatively large demand for reservists[13] suggests that there is a need to assess the joint qualification level of reserve active-status list officers to ensure that the relatively scarce resource of reservists with joint qualifications is managed for the best utilization of those qualifications.

---

[12] Numbers are the combination of officers who agreed and strongly agreed that their current position gives them such experience.

[13] The RC's joint demand is likely to vary widely over time. There is currently a large demand for the reserve component, some of it joint demand, and it may be a long-term demand. The active component has a (more or less) constant force structure with constant joint demands, but the RC is a "surge" force in transition (re-balancing more than the active force) that may continue to be more of an operational force. Even in its more "operational" form, the percent serving on active duty may vary from 25 percent (now), to 10 percent (pre-9/11), to something in between in the future. Analysis within a demand and supply framework should allow for demand and supply (qualification) to vary depending on joint use.

CHAPTER SIX

Conclusions and Recommendations

A framework of law and policy is in place for joint officer management. For active duty list officers, much of that framework is in law; for reserve active-status list officers, most of the framework is in policy. Having a "similar," but not identical, system for reserve joint officer management is a useful construct for reserve active-status list officers because the need for and supply of joint officers and fully joint qualified officers varies by reserve category: traditional unit reservist, active guard/reserve (AGR), and individual mobilization augmentee (IMA). Traditional unit reservists are about 70 percent of the three categories, across all services, at the grades of O-4 and above.

In terms of joint assignments, joint education, joint training, and association with personnel of other services, AGRs have more opportunities for all within the paradigm of their full-time status. Their positions in organizations external to the service are likely to be placed on an eventual Joint Duty Assignment-Reserve (JDA-R), and their need for Joint Professional Military Education (JPME) is facilitated by opportunities at senior service war colleges and Joint Forces Staff College (JFSC). IMAs in active component joint organizations like the Joint Staff (JS), Office of the Secretary of Defense (OSD), and the combatant commands have the advantage of being in a joint billet. They still must manage to obtain first phase of Joint Professional Military Education (JPME I) qualifications and Advanced Joint Professional Military Education (AJPME), but being in a joint billet helps in gaining a JFSC seat. Traditional unit reservists are less likely to serve in

JDA-R billets, and they may also have the disadvantages of lack of time and lack of geographic mobility to overcome if they aspire to joint qualifications.

A strategic approach to joint officer management for reserve active-status list officers must assess the need for officers with prior knowledge, experience, and acculturation before assignment to certain positions. Those positions are not yet identified, nor are the needs for officers with multinational, multiservice, interagency, or intergovernmental knowledge and experience. Determining need should be the first step undertaken. The JDA-R as structured is a list of positions where joint experience might be provided. A comparable database needs to be created for prerequisite joint needs. Recognizing that needs constantly shift and that job descriptions and requisitions should be the basis for matching an officer's qualifications with the need for a particular assignment, this "needs" database must initially be only sufficiently robust to serve as a baseline for how many officer resources must be created (e.g., by service, grade, occupation, location, command) to fill the need on an ongoing basis.

Given identified needs, a strategic approach looks at the current inventory of available officers and projects the future availability given qualification, assignment durations, promotion, and retention rates. A current inventory of available officers does not exist. There are many officers who, because of current deployments and employment, have gained joint training and experience, but these qualifications are not visible within any data system. Populating the JDA-R is a necessary step to determining which officers have received a joint experience, but it is not sufficient to show all officers with joint officer or fully joint qualified status. Reserve active-status list officers serve in many temporary positions or in positions that might not warrant permanent inclusion on a JDA-R. Other means exist to measure and document knowledge and experience and must be used. One must also recognize that the rate of reserve active-status list joint officer qualification should be high now because of mobilization and frequent use. The rate will likely decline in future years, and it will become even more important to measure and document knowledge and experience gained outside a JDA-R time- and billet-based system. Moreover, how officers are managed and assigned into positions affects the overall number of officers who qualify with joint expertise each year. For example, if an IMA serves in a joint organization for 3 years and gains JDA-R qualification, he might continue in that billet, or he might be moved to another billet to allow another officer to qualify. Likewise, if AGRs rotate out of JDA-Rs after 2 years, more officers will have the opportunity to gain joint experience. Alternatively, AGRs could stay in such billets for long tenures and thus develop into a joint "cadre" of well-experienced officers. Modeling and analysis can be used to quantify the tradeoffs of different policies.

Various directives and instructions codify definitions and we placed them into a coherent framework. But such a framework begs the question as to why certain distinctions are made. If second phase of Joint Professional Military Education (JPME II) and AJPME are "similar" and cover the same learning objectives, why the distinction between them? If there is no distinction, why not permit active duty list officers to complete AJPME? If "joint specialty officer" (JSO) and "fully joint qualified" (FJQ) both mean an officer is a Joint Qualified Officer, why the distinctions? If the dominant characteristic of a JSO nominee is that he has completed JPME II but not yet a Joint Duty Assignment List (JDAL) assignment, how is he different from a "Joint Officer" as defined in the DoD Instruction for reserve joint officer management? The concept of valid qualification should matter more than the semantic labels. The joint officer system could either manage the semantic differences or simplify them into a common lexicon. Neither changes the concept of officers who are qualified by knowledge and/or experience in joint matters.

Recommendations

We recommend that DoD adopt a strategic approach for reserve active-status list officers.

- First, data are needed about requirements and should be collected. The services should be directed to incorporate data about the needs for prior joint education, or prior joint experience, and thus for fully joint qualified officers and joint officers into their manpower databases. At the same time, the JS should implement and maintain a stand-alone database for at least the JS, OSD, combatant commands, and other external organizations until such time that the manpower systems are changed. A one-time data collection[1] should be done to populate this database initially with updates as needed. In particular, the latter two sources of demand could vary most for the reserve component over time as they surge to meet changing geostrategic situations.

- Second, a supply-oriented database such as the JDA-R should also be populated with positions that provide joint experience. Procedures for doing this for the JDA-R are laid out in the DoD Instruction and should be implemented.

- Third, OSD, the JS, and the services should specify policies and procedures for capturing information about qualifying knowledge and the experience of officers beyond that which will be captured by the billet- and time-based JDA-R system. We recommend that a point system or an accomplishment record approach be used. Officers who aspire to be joint officers or fully joint qualified would document their record using letters of evaluation, officer efficiency reports, awards, recommendations, award of OSD or Joint Staff or other relevant badges, assignment orders, civilian experience, and documentation of frequent, recent, intensive but short periods of training and assignments, etc., which would be reviewed by a senior officer or a formal board. The administrative burden would be reduced by making this an opt-in policy that focuses on joint development and the use of officers who want such development either as a professional aspiration or as a needed step to progressing along a career path.

---

[1] This process is outlined in Appendices B and C of *Framing a Strategic Approach for Joint Officer Management* (MG-306-OSD), and a similar process could be followed.

- Fourth, the future supply of qualified reserve active-status list officers should be projected using modeling of JFSC AJPME seats, JDA-R positions, assignment duration, qualification by other means, and likely promotion and retention rates.
- Both current and projected inventory needs to be compared to the demand to determine where shortages and overages exist as a basis for formulating appropriate policy alternatives. This analysis would determine the extent to which the need for officers with joint education and experience can be satisfied by the number of qualifying billets, other qualifying experiences, and educational seats combined with the use, promotion, and other management practices for officers of different reserve categories and occupational communities.[2] The strategic plan should lastly determine the policies and practices to align the amount of jointness available with the demand for jointness.
- Finally, the implemented strategic approach, which recognizes both the need for jointness among reserve active-status list officers and the complementary means to acknowledge and accredit joint qualifications for reserve officers, should be evaluated and considered for its application to active duty list officers.

---

[2] Because of continuing change in the roles and missions of the reserve components, these assessments will need to be made periodically, if not continually, to avoid long-term management problems.

Availability of Current Data to Support Reserve Component Joint Officer Management

This appendix examines the availability and appropriateness of existing data for determining the jointness of current reserve active-status list officers, which was one of the tasks we were asked to accomplish. We first examined the kinds of data tracked by the Reserve Component Common Personnel Data System (RCCPDS), which is the "official source to provide statistical tabulation of Reserve component strengths and related data for use throughout the Department of Defense," according to DoD Directive 1205.17. Because the data on the RCCPDS are limited at best and not well suited to supporting the goals and objectives of RC JOM without certain alterations, we also examined the data collected and reported by the Air Force on the Air Force Reserve (AFR) and the Air National Guard (ANG), to see how well these data could be used to track joint education, joint assignments, and career progression of reserve component officers. This appendix serves as a basis for understanding our recommendations for data collection.

Reserve Component Common Personnel Data System

As mentioned above, RCCPDS was established as the official database for the reserve components and is operated and maintained by Defense Manpower Data Center. RCCPDS is cited as the official record and appears to provide the only systemic assessment of the reserve force from a corporate perspective that also allows trend analysis. This process creates a common convention and framework through which the community interacts. Under Secretary of Defense for Personnel and Readiness (USD(P&R)) and Assistant Secretary of Defense for Reserve Affairs (ASD(RA)) are tasked with providing guidance on policy, quality, and usage; Deputy Under Secretary of Defense for Program Integration (DUSD(PI)) is responsible for collection and coordination with the services. Each of the service reserve organizations submits personnel data compiled according to this directive and the standards defined in DoD Instruction 7730.54.

DoDD 1205.17 states that RCCPDS must be "consistent with the active force database" and requires that certain data elements be maintained for each reserve member, including "physical condition, dependency status, military qualifications, civilian occupational skills [and] availability for service." DoD Instruction 7730.54 provides guidance on the levels of desired precision for various data categories in the RCCPDS: 100 percent accuracy on reserve category status, transaction and person identifiers; 98 percent accuracy for demographic information, appointment and activation status, military grade and pay, armed forces qualification test scores, and unit identification codes; and 95 percent accuracy for all other fields. In total, more than 139 fields are recorded in this dataset.

The RCCPDS data system appears to be widely used for administrative and statistical reporting on the reserve components, but it is clear that several additions and modifications would need to be made if RCCPDS is to be used as a strategic resource for RC JOM. As it stands, there is some information on joint backgrounds and/or joint education, and some fields that might be relevant for RC JOM are presented in Table A.1. However, many of these fields are either limited or incomplete in scope or lost through poor formatting and data integrity. For example, neither professional military education nor experience-related variables that would be relevant are emphasized by the instruction's top-level procedures (as compared to language proficiency and command status). Nor is it geared to tracking experiences gained through various assignments or representing an officer's development path.

There are three types of structural issues that hinder the applicability and usefulness of RCCPDS for RC JOM. These include concerns with how data are gathered, structural relationships, and lower-level operational issues.

Table A.1
Data Elements on the Reserve Component Common Personnel Data System Potentially Useful for Reserve Component Joint Officer Management

Field	Data Item
1	Reserve Component
1	Military Service (Uniformed Service Organization Code)
1	Service Component (Uniformed Service Organization Component Type Code)
2	Reserve Component Category (RCC) Designators
2	RCC Designators
2	Training and/or Retirement Category (TRC) Designators
3	Military Technician, Active Guard and Reserve, or Full-Time National Guard Duty Statute Identifier
4	Key Employees
5	IRR Drilling Status 20
5	IRR Drilling Status Indicator
5	Filler
10	Sex (Sex Category Code)
11	Marital Status (Person Marital Status Code)
12	Race and/or Population Group (Race Code)
13	Ethnic Group
13	Ethnic Group Code
14	Faith Group Code
15	U.S. Citizenship Status
15	U.S. Citizenship Status Code (Person Organization Person Role Code)
15	Filler
16	U.S. Citizenship Origin Code (Citizen Citizenship Origin Code)
17	Disputed Record Indicator
18	Education Designator
18	Education Designator Code
18	Filler
26	Date of Initial Entry into Uniformed Service (DIEUS)
27	Date of Initial Entry into Reserve Forces (DIERF)
28	Pay Entry Base Date (PEBD)
30	Source of Initial Commission and/or Appointment
30	Source of Initial Commission for a Commissioned Officer
30	Source of Initial Appointment for a Warrant Officer
31	Initial Appointment Date
31	Date of Initial Appointment for a Commissioned Officer
31	Date of Initial Appointment for a Warrant Officer
31	Date of Initial Appointment for a Commissioned Warrant Officer
32	Prior Service Status Indicator (Regular)
33	Prior Service Status Indicator (Selected Reserve)
34	Length (Years) of Current Selected Reserve Agreement and/or Service Commitment

Table A.1—Continued

35	Active Duty Start Date
36	Active Duty Stop Date
37	Date of Expiration of Enlistment in the Ready Reserve
38	Effective Date of Current Enlistment, Reenlistment, or Extension of Enlistment Agreement
39	Date of Expiration of Selected Reserve Obligation
40	Date of Rank
41	Pay Grade, Uniformed Services
41	Pay Grade, Uniformed Services Code (Pay Plan Code)
41	Pay Plan Grade (Pay Plan Grade Ordinal Identifier)
42	Total Days Active Federal Military Service
43	Date of Expiration of Statutory Military Service Obligation (MSO)
46	Service Occupation Code (Primary)
47	Service Occupation Code (Secondary)
48	Basic Branch or Specialty (Officer Only)
49	Professional Military Education Level
49	Basic Professional Military Education Level Code
49	Joint Professional Military Education Level Code
49	Joint Professional Military Education Completion Date
50	Command Status of Commissioned Officer
51	Armed Forces Qualification Test (AFQT) Percentile Score (Enlisted Only)
52	Date Assigned
52	Standby Reserve
52	Retired Reserve
63	Assigned Military Unit Designator (Unit Identification Code (UIC))
64	Assigned Unit Location
65	Duty Military Unit Designator Unit Identification Code (UIC)
66	Service Occupation Code (Duty)

First, the fields required by RCCPDS are generally insufficient in number, quality, and detail from the point of view of supporting a strategic joint officer management system. JPME (as well as professional military education [PME]) proficiency is tracked only for the "highest level . . . completed by an officer" and does not allow tracking over time. The associated "JPME Level Completion Date" field is also not significantly populated. Some indicators of experiential occupational and unit designators (though no billets) do appear in the data and could be used as supplemental measures indicating joint experience. However, the data do not provide significant insight into the strategic joint capabilities of the officer corps and are insufficient other than as a partial headcount.

Second, RCCPDS is dependent on the services for monitoring the quality and completeness of the data they provide and for ensuring that the data comply with the standards laid out by DoDI 7730.54. RCCPDS only manages the integration and reformatting of the data. Thus, each submitting organization must be individually contacted if the issue is not captured in the instruction or if data are inconsistent or missing. The issue of data completeness is further complicated by the fact that different organizations may have individual responsibility for the integrity of certain fields across officer records. For example, the joint/service schools that grant PME or JPME credit may provide data separately from the standard personnel management operation. Thus, integration of data may occur at multiple points in the process and require further identification.

Third, operational issues associated with internal data quality or preparation of the master and transaction files may affect the availability of data for use by other agencies. For example, both the unit identification (UIC) and "Prior Service Status Indicator" fields are collected and populated but are subject to limitations for public extract. "Prior Service Status Indicator" might be a useful measure for determining whether experiences may have been gained as an active duty service member but is unavailable due to internal quality control problems. Fields recorded in the data set may not be available for all reservists. For example, Assigned and Duty UICs are available only for the Selected Reserves. In addition, because each record is structured as a "flat file" with fixed fields, attribute information is overwritten when it changes. Thus on most counts, RCCPDS does not provide a longitudinal history of an officer's career.

Data from September 2005 Reserve Component Common Personnel Data System

We used data from the September 2005 RCCPDS to examine the level of missing information in variables that would be relevant to RC JOM. Reserve officers who have the potential to become joint qualified are likely to be in grades O-4 and above and in the Selected Reserve. As of September 2005, there were 114,055 officers in the Selected Reserve, of whom 67,902 were in grades O-4 and above. The distribution of these officers by reserve component is shown in Table A.2.

Table A.2
Distribution of Officers in Grades 0-4 and Above by Reserve Component

Reserve Component	Percentage (n=67,902)
Army Reserve	29.1
Army National Guard	17.6
Naval Reserve	18.6
Air Force Reserve	18.0
Air National Guard	13.1
Marine Corps Reserve	3.7

Because it is likely that demand for joint officers will differ among officers by reserve category, Table A.3 shows the number of officers distributed by reserve component and reserve category. We selected only those who were (a) unit reservists or drilling reservists,[1] (b) Individual Mobilization Augmentees (IMAs), or (c) Active Guard/Reserve (AGRs) (n = 67,754).[2]

The education variables are of special interest for purposes of tracking joint officer development. Table A.4 shows the distribution of officers by reserve component and professional military education. This variable seems to be largely complete, although a not insubstantial number of officers seem not to have had any PME. Of note are the 6,864 officers in the Naval Reserve who have value of "none" and 2,494 coded as "unknown." The Air Force Reserve and the Air National Guard report "none" for 1,081 and 1,423 officers, respectively.

Table A.5 shows the data recorded for JPME for these officers. The RCCPDS records only the highest level of JPME credit received, despite the precision and volume of other data elements in the system. As is clear from the table, the components are inconsistent with how they record this information. The Army Guard, for example, does not

---

[1] In this group, following RCCPDS, we include a small number of full-time members who are in a special category (category V). There were only 145 of these reservists as of September 2005.

[2] This excludes officers who were in the training pipeline/other training programs such as chaplain or medical (category X) or on initial active duty for training (category F).

Table A.3
Distribution of Officers in Grades 0-4 and Above by Reserve Component
and Reserve Category

Reserve Component	Drilling Unit Member	Active Guard/ Reserve (AGR)	Individual Mobilization Augmentee (IMA)	Total
Army Reserve	14,323	2,635	2,781	19,739
Army National Guard	8,936	2,871	0	11,807
Naval Reserve	10,976	1,455	208	12,639
Air Force Reserve	5,942	678	5,575	12,195
Air National Guard	7,056	1,832	0	8,888
Marine Corps Reserve	1,200	261	1,025	2,486
Total	48,433	9,732	9,589	67,754

appear to record data on JPME; the Army Reserve seems to have more complete data, with no "unknowns." All the other components have a substantial number of "unknowns."

RCCPDS also reports JPME completion date. At most, one would expect to see records for about 12,597 cases based on Table A.5. As shown in Table A.6, there are differences by component—the Air Force Reserve, Air National Guard, and Naval Reserve appear to record this with some consistency, the Army Reserve is missing information for about 1,000 cases, and the Marine Corps Reserve on about 150 cases.

Whereas JPME is separately identified (although with its own problems), joint experience appears to be only partially accessible and then based only on other attributes. This makes it difficult to operationalize because of concerns with the quality and comprehensiveness of the data. Core and duty military occupational specialty (MOS) attributes appear to be the only fields that would capture general job skill characteristics, as no billet or assignment attributes are required for submission to RCCPDS. Although these fields are standardized by record position and service format, the inclusion of the skill qualifiers is optional.

Table A.4
Distribution of Officers in Grades 0-4 and Above by Reserve Component and Professional Military Education

Reserve Component	Senior Service School	Intermediate Service School	Skill Progression School	Initial Skill	None	Unknown
Army Reserve	904	9,406	7,725	1,381	229	95
Army National Guard	778	6,073	4,575	449	61	0
Naval Reserve	44	74	447	2,723	6,864	2,494
Air Force Reserve	165	446	10,503	0	1,081	0
Air National Guard	111	373	6,992	0	1,423	0
Marine Corps Reserve	40	219	153	2,064	10	0
Total	2,042	16,591	30,395	6,617	9,668	2,589

Table A.5
Distribution of Officers in Grades 0-4 and Above by Reserve Component
and Joint Professional Military Education

Reserve Component	Advanced	Initial	None	Unknown
Army Reserve	4	1,420	18,316	0
Army National Guard	0	0	0	11,936
Naval Reserve	18	83	7,465	5,080
Air Force Reserve	7	6,170	0	6,018
Air National Guard	8	4,737	0	4,154
Marine Corps Reserve	1	149	0	2,336
Total	38	12,559	25,781	29,524

Table A.6
Distribution of Officers in Grades 0-4 and Above by Reserve Component
and Joint Professional Military Education Completion Date

Reserve Component	Advanced	Initial	None	Unknown
Army Reserve	4	1,420	18,316	0
Army National Guard	0	0	0	11,936
Naval Reserve	18	83	7,465	5,080
Air Force Reserve	7	6,170	0	6,018
Air National Guard	8	4,737	0	4,154
Marine Corps Reserve	1	149	0	2,336
Total	38	12,559	25,781	29,524

Overall, the current format of RCCPDS does not lend itself to longitudinally tracking an officer's development, because fields are overwritten as they are updated.

We now turn to data collected and reported by the Air Force Reserve and the Air National Guard to examine the feasibility of these data being used to support reserve component joint officer management.

Data on Air Force Reserve and Air National Guard

AFR collects considerably more data than it submits to the RCCPDS. AFR personnel management have been operating under an internal, legacy "PDS" Personnel Data System that was custom-developed and that has evolved organically (both as a system and as a record structure) over the last 40 years.[3] The Military Personnel Data System (MILPDS) is the traditional transactional accounting system that covers promotions, assignments, duty, and other personnel information.[4] These systems are undergoing modernization, but in the process, some historical data appear to have been lost. As a result, historical data on officers are not available from internal sources and the only source is data that were submitted to RCCPDS.

Uniform Officer Record (UOR) files are maintained by the Air Force Military Personnel Center and reported to outside parties on a yearly basis as SAS datasets (although monthly, weekly builds are maintained internally). UOR also features transactional accounting for individual officer records. MILPDS serves as the underlying framework that enables the UOR and all other personnel applications.[5]

The UOR strength file, maintained by the Air Force Manpower and Personnel Center (AFMPC) at Randolph Air Force Base, Texas, contains a record for each officer who is either currently on active duty or projected to be gained to active duty in the Air Force. The UOR Gain/Loss file contains accessions, reenlistments, separations, retirements, promotions and demotions, permanent change of station, and extensions. Along with demographic variables (sex, race/ethnicity, date of birth, marital status, etc.), the file contains data on grade, duty status, service dates, assignments, aeronautical qualification, and education. Each fiscal year file also contains information on the officer's 10 previous assignments. Files are as of end date of a fiscal year, usually September 30.

The reserve files (split into reserve and guard) contain more information and detail than the data reported to RCCPDS. The Reserve file contains data on 54,969 AFR officers (both active and inactive reserve), and the Guard file contains data on 13,166 ANG officers

---

[3] United States Air Force Armstrong Laboratory, Summary of Data Bases, December 1992, http://www.icodap.org/040510/SUMMARY1992.htm#n.%20UOR (as of March 2006).

[4] United States Air Force Personnel Center, Military Modernization, http://www.afpc.randolph.af.mil/dlearn/milmod/default.htm (as of March 2006).

[5] Draft Air Force Personnel Information Systems Concept of Operations (as of July 14, 2004), http://www.icodap.org/HRRD/CONOPS-HRRD.htm (as of March 2006).

(active ANG). The Guard file encompasses almost 1,500 attribute fields, which is considerably higher than the 900 fields in the Reserve file. Many of the common attributes do appear relevant for RC JOM, but because of lack of full documentation, it is unclear how useful the remaining fields are likely to be. It is possible to track changes over time for many of the significant attributes because later data are maintained as separate fields.

Table A.7 demonstrates the breadth of included field types, with history (prior 3 occurrences or prior 24 as an example), captured by the standard dataset. Fields of particular interest are highlighted—

Table A.7
List of Variables on the Air Force Reserve and Air National Guard Files That Could Be Used to Support Reserve Component Joint Officer Management

Labeled Fields With Prior History	Descriptive Category (Rand)	National Guard		Reserves	
		Prior 3	Prior 24	Prior 3	Prior 24
ACAD EDUC LEVEL MET	Education - Academic	x		x	
ACAD INST NAME HIGH	Education - Academic	x		x	
ACAD VOC EDUC LEVEL HIGH	Education - Academic	x		x	
ACAD VOC EDUC LV YYMM HIGH	Education - Academic	x		x	
ACADEMIC SPECIALTY HIGH	Education - Academic	x		x	
AFIT ED LEVEL PROG	Education - Afit	x			
AIRCRAFT MOST RECENT	Rating	x			
AIRCRAFT MOST RECENT FLOWN	Rating	x			
AIRCRAFT MOST RECENT HOURS FLOWN	Rating	x			
ASG AVAIL - YYMM	Assignment			x	
ASG AVAIL CODE	Assignment			x	
ASG LIMIT	Assignment	x		x	
ASG LIMIT EXP - YYMM	Assignment	x		x	
BLOCK ASG EXPR DT - YYMM	Assignment	x			

Table A.7—continued

BLOCKED ASSIGNED	Assignment	x			
DECOR AUTH	Decoration	x		x	
DECOR CLOSE DATE	Decoration	x		x	
DECOR CONDITION	Decoration	x		x	
DECOR HQ	Decoration	x		x	
DECOR NR	Decoration	x		x	
DECOR ORDER - YYMM	Decoration	x		x	
DEPN CHILD HSHLD SEX	Personal			x	
DEPN CHILD HSHLD YOB	Personal			x	
HIST ACQ POSN CAT	Acquisition - History	x	x		
HISTORY ACQ POSITION INDICATOR	Acquisition - History	x	x		
HISTORY COUNTRY/ STATE	Duty - History	x	x		
HISTORY DAFSC	Duty - History	x	x	x	x
HISTORY DUTY COMMAND LEVEL	Duty - History	x	x		
HISTORY DUTY EFF DATE	Duty - History	x	x		
HISTORY DUTY LOCATION	Duty - History	x	x		
HISTORY DUTY SPEC EXP ID	Duty - History	x	x		
HISTORY DUTY TITLE	Duty - History	x	x		
HISTORY FUNCTIONAL ACCOUNT	Duty - History	x	x		
HISTORY JDA TR MPWR RMKS	Duty - Joint	x	x		
HISTORY MAJCOM ID	Duty - History	x	x		
HISTORY OPR ADRS	Duty - History			x	x
HISTORY OPR CONTROL	Duty - History			x	
HISTORY ORGN DET NUMBER	Duty - History	x	x		
HISTORY ORGN KIND	Duty - History	x	x	x	x
HISTORY ORGN NUMBER	Duty - History	x	x	x	x
HISTORY ORGN TYPE	Duty - History	x	x	x	x
HISTORY PERF INDIC	Opr	x		x	
LANG PRO PAY EFF DT	Language	x			

Table A.7—continued

LANG SELF ASSESS ID	Language	x		x	
LANG SELF ASSESS LVL	Language	x		x	
LANGUAGE ID	Language	x		x	
LANGUAGE LISTEN COMPRE	Language	x		x	
LANGUAGE READ COMPRE	Language	x		x	
LANGUAGE SELF ASSESSED DT	Language	x		x	
LANGUAGE SPEAK COMPRE	Language	x		x	
LANGUAGE TEST DT - YYMM	Language	x		x	
MILITARY DECOR AWARD	Decoration	x		x	
PAL	Unknown			x	
PROF MIL COURSE	Education - Military	x		x	
PROF MIL MET STY SCH	Education - Military	x		x	
PROF MIL SCH YR – YYMM	Education - Military	x		x	
PROF SPEC COURSE	Education - Professional	x	x	x	x
PROF SPEC COURSE DT - YYMM	Education - Professional	x	x	x	x
PROF SPEC COURSE RSN	Education - Professional	x	x	x	x
PROJ PROF SPEC COURSE	Education - Professional	x		x	x
PROJ PROF SPEC COURSE GRAD DATE	Education - Professional	x		x	x
PROJ PROF SPEC COURSE START DATE	Education - Professional	x		x	x

categories tracked include education, skill ratings, personnel specialties/ AFSC and assignments—experiences, particularly JDA. This is much richer than what RCCPDS records, with more detail and explicit labeling. Of particular note is that start and graduation dates are captured for all PME, as well as projected PME.

Assuming that the field quality/integrity can be ensured, it seems like this database could be used to support RC JOM. Education and

Experience metrics are clearly structured in the USAFR's UOR data fields. As an example, over 48,000 of the 55,000 officers have at least their first "Professional Spec Course" field populated by a course catalog code, whereas 3,970 have the previous 10 courses recorded and maintained by the same level of detail. Additional fields that might be of interest would be the Course_ID, Course_Title and Class_Start_Date fields, and duty and other accounting/status variables. It is difficult to make an assessment of the latter fields because some of them are not labeled and the descriptions are not very informative.

Overall, the Air National Guard reports less information on the active history of its officers than does the USAFR.

Issues to Consider

An internal issue exists with how the USAFR, versus the active duty USAF, recognizes (J)PME credit. The discrepancy is based on opportunity and equity: JPME I credit is granted along with PME after attending intermediate service school programs, but JPME II and AJPME are treated very differently. Credit for AJPME is not recorded on the PME table but instead treated as a secondary technical skill rather than recorded under education. One reason for this is that given the fewer opportunities available to reservists for taking the course, AJPME credit is seen as an unfair promotion discriminator. Thus, any potential crosswalk between the active duty and reserve/guard files needs to take this into account.

A second issue is that because JPME is now tracked separately from PME, there may be double counting and inconsistency in recording. In addition, it is important to ensure consistency among organizations of how credit is given and recorded.

Theory Behind Training and Experience Evaluation Methods

Training and experience rating methods are used to predict future job performance through evaluations of resumes, applications, and/or other documents provided by applicants. Methods are based on assumptions of positive correlations between education and experience, and job performance.[1] Experts in human resource management have reasoned that the amount and quality of education are indirect measures of knowledge, skills, and abilities (KSAs), and these KSAs are correlated with job performance.[2] Testing of the framework, however, has resulted in relatively low validities.[3]

Further research shows that two significant moderators of the correlation between experience and job performance are *length of experience* and *cognitive complexity* of the job. Although the correlation between

---

[1] These methods are not the only means of personal selection. Other frequently used methods include biodata questionnaires and cognitive tests to measure aptitudes in various areas. However, these methods are not relevant for this study.

[2] Schmidt, Caplan, et al., 1979, cited in Michael A. McDaniel, Frank L. Schmidt, and John E. Hunter, "A Meta-Analysis of the Validity of Methods for Rating Training and Experience in Personnel Selection," *Personnel Psychology,* Vol. 41, No. 2, 1988, p. 284.

[3] According to McDaniel et al., 1988, ratings seldom correlate more than .40 with KSAs, and KSAs seldom correlate more than about .50 with job performance. Thus the final validity coefficient of a traditional rating is estimated to be about .40 × .50, or .20. Validity coefficients below .11 are unlikely to be useful measures. Coefficients between .11 and .20 are sometimes useful. Coefficients between .21 and .35 are likely to be useful and coefficients above .35 are considered to be very beneficial. Taken from "Testing and Assessment: An Employer's Guide to Good Practices," U.S. Department of Labor Employment and Training Administration, 3/99, pp. 3–10.

job experience and job performance is always positive, there are diminishing returns for length of experience (Figure B.1). In other words,
after a certain amount of experience, further experience has less of an
impact on job performance. A study on the job experience correlates of
job performance found a much higher correlation between experience
and job performance for samples with mean job experience of 3 years
or fewer as compared to a similar sample with a mean job experience
of 12 years or greater.[4] In the former, the correlation was .49, but the
latter was only .15. One explanation offered for the sharp drop in the
relationship is that early career experience yields the greatest improve-

Figure B.1
**Relationship Between Length of Experience and Job Performance (Based
on Tim McGonigle, Theory Behind Training and Experience Measures, 2003)**

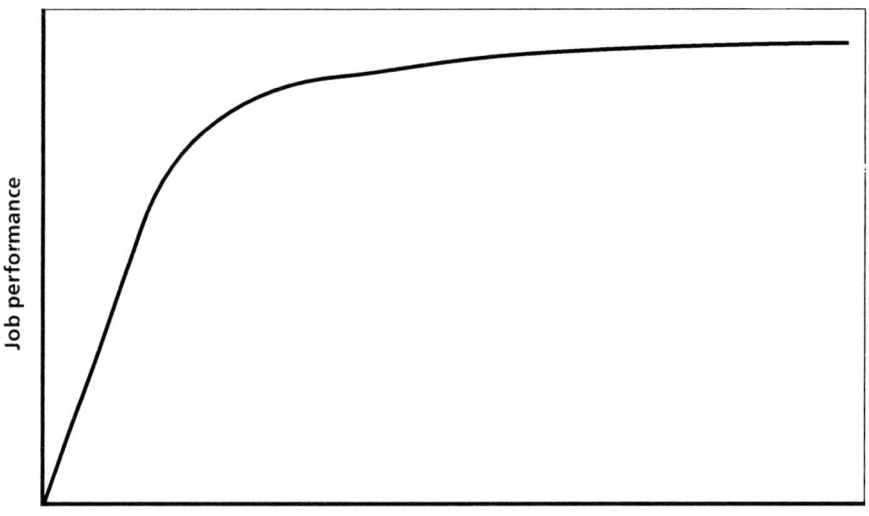

[4] Michael A. Mc Daniel, Frank L. Schmidt, and John E. Hunter, "Job Experience Correlates
of Job Performance," Journal of Applied Psychology, Vol. 73, No. 2, 1988, p. 330.

ments in job knowledge. Experts reason that as the knowledge of the employee expands, increasing amounts of experience add smaller and smaller increments to knowledge.[5]

A second moderating factor between experience and job performance is the *complexity* of the job at hand. The correlation between experience and performance is found to be higher for low-complexity jobs than for high-complexity jobs. In the same study, for low-complexity jobs, the correlation was .39, whereas the correlation for high-complexity jobs was .28. Thus for lower complexity jobs, experience is more important to job performance.[6]

The Schmidt-Hunter model may explain the moderating effect of job complexity on the relationship between experience and performance. The model holds that experience impacts job performance most through the development of job knowledge.[7] High-complexity jobs often require formal education, and this formal education increases job knowledge before working on the job. More specifically, prework education targeted to specific job-related knowledge should cause one to achieve a high level of job performance more quickly. In low-complexity jobs, job experience is often the sole source of job knowledge. As it relates to joint officer training, joint assignments can be considered high-complexity jobs. Reserves that complete education, particularly targeted joint education such as AJPME before a joint billet, will have amassed job knowledge and require less experience to reach high levels of performance.

Another moderating factor of job experience and performance is the *stability of the job knowledge base* required for a job. For jobs where the knowledge base changes rapidly, past experience and knowledge gained from it will have limited effect on job performance.[8]

Recency and *frequency* are two additional variables that impact the correlation between experience and job performance. Recency refers to

---

[5] Tim McGonigle, "Theory Behind Training and Experience Measures," presented at IPMAAC Conference, Baltimore, MD, June 2003.

[6] McDaniel, et al., "Job Experience Correlates of Job Performance," p. 329.

[7] Ibid, p. 330.

[8] Tim McGonigle

the length of time since the applicant's experience. Studies show that more recent experience has a stronger correlation with job performance than experience further in the past, leading those with recent experience to gain proficiency faster. This is explained by the fact that task proficiency may diminish over periods of non-use, making past experience less relevant. An issue that should be considered for those with non-recent experience is how long and intense a retraining period would need to be to overcome the gap in experience. More frequent experience also has a stronger correlation with job performance than less frequent experience.[9]

Validity and How Experience Is Measured

The correlation between experience and job performance also depends on the way "experience" is constructed and measured. Most studies have used time on the job, or tenure, to measure work experience. A few studies have measured experience by counting the number of times an individual performs a given task.[10] Other approaches have focused on the actual content of the experience. There are two general dimensions that capture the various measures of work experience: *measurement mode* and *level of specificity*.[11] Measurement modes include three types of measures: (a) time-based measures that refer to job or organizational tenures, that is, months or years on a job; (b) amount-based measures that refer to numerical counts such as the number of times a task was performed or the number of different jobs in an organization;

---

[9] Shelly Butler, "Other Variables Affecting T&E Measures," presented at IPMAAC Conference, Baltimore, MD, June 2003.

[10] Empirical analysis by Philip Djang et al. ("The Army's Unit Training Model," in *Final Report, 66th MORS Symposium*, Alexandria, VA: Military Operations Research Society, 1998) determined that, on average across tasks, additional repetitions beyond eight yield little to no additional increase in performance for all training methods. See also John F. Schank, Harry J. Thie, et al., *Finding the Right Balance: Simulator and Live Training for Navy Units*, Santa Monica, Calif.: RAND Corporation, MR-1441-NAVY, 2002.

[11] Miguel A. Quinones, J. Kevin Ford, and Mark S. Teachout, "The Relationship Between Work Experience and Job Performance: A Conceptual and Meta-Analytic Review," *Personnel Psychology*, Vol. 48, No. 4, 1995, p. 891.

and (c) type-based measures that categorize experience qualitatively. Each of these three modes can be operationalized at three levels of specificity. These are task, job, and organizational levels (see Table B.1). The 2005 Joint Officer Management Census survey polled officers serving in billets that were likely to require joint experience or joint education or provide such experience. More than 21,000 survey responses were collected. The survey question included many of the measures shown in Table B.1.[12] These measures will be analyzed in a forthcoming RAND report.

As shown in Table B.1, at the task level, experience can be measured in number of times performing the task (amount), the types of tasks they have performed (type), and the amount of time spent working on a given task (time). At the job level, individuals can differ in the number of total jobs they have held (amount). They can differ in the experience performing different types of jobs that vary in terms of prestige, difficulty, or criticality (type). Individuals can also differ in the amount of time spent in a particular job (time). Lastly, experience can be measured at the organizational level. Individuals can differ in the number of organizations for which they have worked (amount). They can vary in the type of organization a person has worked (type). They can differ in the amount of time spent in a given organization (time).[13]

The strongest correlation between experience and job performance is seen when the measurement mode is *amount* and the level of specificity is *task*. A study found that when experience is measured using the *amount* measurement mode, the correlation with job performance is .43 as compared to *time* and *type* at .27 and .21, respectively.[14] When experience is measured using the *task* level of specificity, the cor-

---

[12] Sheila Nataraj Kirby, Al Crego, Harry J. Thie, Margaret C. Harrell, Kimberly Curry, and Michael S. Tseng, *Who Is "Joint"? New Evidence from the 2005 Joint Officer Management Census Survey*, Santa Monica, Calif.: RAND Corporation, TR-349-OSD, 2006.

[13] Quinones, et al., The Relationship Between Work Experience and Job Performance: A Conceptual and Meta-Analytic Review," pp. 892–893.

[14] Ibid, 903.

Table B.1
A Conceptual Framework of Work Experience Measures

		Measurement Mode		
		Amount	Time	Type
Level of specificity	Organization	Number of organizations (JS, COCOM, OSD, services, etc.)	Tenure in organization	Nature of organization (e.g., service, functional, external to service)
	Job	Number of jobs or aggregate number of tasks	Tenure in job	Job complexity (e.g., strategic, operational, tactical)
	Task	Number of times performing a task	Time on task	Task difficulty, complexity, or importance

SOURCE: Adapted by authors from Quinones et al., "The Relationship Between Work Experience and Job Performance"

relation with job performance is .34 compared to *job* and *organization* level at .22 and .16.[15] Experience measured by *amount* at the *task* level best indicates what individuals actually do on the job.

We have labeled the JDAL construct as a time- and billet-based system for evaluating job experience. Using the typology of Table B.1, it uses the measurement mode of time and the level of specificity of job.

Evaluation of Education and Experience

A variety of methods are used in personnel selection that could potentially be used in certifying "joint" officers. These methods differ in terms of their validities, appropriateness for job types (entry level, low-skill versus upper-level, high-complexity jobs), and cost and difficulty of administration. Ironically, the most commonly used measurement methods, holistic judgment and the traditional point method, are shown to be the least valid. Less commonly used methods such as

---

[15] Ibid.

KSA-oriented methods and accomplishment records appear to have greater validities.

As a general rule, evaluation methods should incorporate job analysis because the basis of evaluation is matching backgrounds of applicants to job requirements. Job analysis is a process that determines (a) the reason for the job, (b) job duties that are critical or fundamental to the performance of the job, (c) job setting or the work station and conditions where the essential functions are performed, and (d) job qualifications or the minimal skills an individual must possess to perform the essential functions.[16] Standardization should also be incorporated into the evaluation method. Standardization may include standardization of questionnaires, rating forms, and the rating process through reliance on training programs for raters or protocols of rating procedures that serve as job aids.

This section describes different measurement methods, respective advantages and disadvantages, and methodologies that may improve the use of a specific measurement method.

Holistic Judgment

Holistic judgment is the most frequently used method to evaluate training and work experience, but it is not a formally scored measurement device.[17] It generally consists of an unstructured review of a resume or application. Because of its unstructured nature, holistic judgment is difficult to evaluate for validity and reliability. The criteria and process for making judgments about applicants from resume data generally exist only in the minds of individual evaluators, and thus are both unrecorded and unstandardized.

---

[16] U.S. Department of Labor. See http://www.dol.gov/odep/pubs/fact/analysis.htm

[17] Ronald A. Ash, James C. Johnson, Edward L. Levine, and Michael A. McDaniel, "Job Applicant Training and Work Experience Evaluation in Personnel Selection," *Research in Personnel and Human Resources Management*, Vol. 7, 1989, p. 199.

Minimum Qualifications

Minimum qualifications (MQs) (e.g., time in certain billets, completion of certain educational courses) is the method currently used by the military to certify active-duty as "joint" officers. MQs are statements of education, experience, and/or closely related personal attributes needed to perform a job satisfactorily that are used as standards to screen applicants.[18] Generally, MQs serve as the first hurdle in a selection process and thus critically affect the entire process. They are minimal in the sense that they are intended to screen out applicants who are unable to perform the job and to screen in applicants who are able to perform at a minimally acceptable, yet satisfactory, standard.

The "qualifications" sought by employers are typically expressed in one of three ways: (a) as a minimum amount and kind of education and experience, (b) as a preferred amount and kind of education and experience, or (c) as a statement of the job competencies the organization seeks, regardless of how these might have been acquired.[19]

A semistructured review of a resume or application is used to identify the minimum qualifications.

In the case of certifying active-duty officers as "joint" qualified, MQs are a minimum amount and kind of education and training. These include completion of JPME II and 2 years served in designated joint assignments for certain occupations and 3 years for other field grade officers. MQs often allow substitution of experience for education and vice versa. This study seeks to determine whether the military could allow substitutions of greater training and education for experience for reserve officers to be considered joint. The military currently allows substitution of experience for education in designating some joint specialty officers (JSOs).

A disadvantage of using MQs for employee selection or job certification is that it is often difficult to equate education and experi-

---

[18] Edward L. Levine, Doris M. Maye, Ronald A. Ulm, and Thomas R. Gordon, "A Methodology for Developing and Validating Minimum Qualifications (MQs)," *Personnel Psychology,* Vol. 50, No. 4, 1997, pp. 1009–1024.

[19] Ash et al., "Job Applicant Training and Work Experience Evaluation in Personnel Selection."

ence across applicants. For example, for joint officers, factors such as the individual joint school, coursework, and instructors weigh into the educational experience. Joint duty assignments (JDAs) across two candidates are equally or more diverse. JDAs may differ in terms of the interaction required with other services or the areas of joint knowledge that are applied. Despite frequent use of MQs, meager research has been conducted on the subject, and its validity is unknown.[20]

A recent study by Levine et al. tested a new methodology for developing and validating MQs that was able to produce high levels of content validity. This methodology involves a new type of job analysis with scales to determine MQ domains and validate the constructed MQ profiles. The first step of the job analysis is research and observation that leads to the preparation of draft lists of tasks and KSAs for the particular job. Subject matter experts (SMEs) then use scales to evaluate tasks and KSAs for their impact and relevance in establishing MQs.

Examples of scales[21] applied to tasks for defining domains of MQs are:

- **Perform at Entry:** Should a newly hired employee be able to perform this task immediately or after a brief orientation/training period? (Yes/No)
- **Barely Acceptable:** Must even barely acceptable employees be able to perform this task correctly with normal supervision? (Yes/No)
- **Importance of Correct Performance:** How important is it for this task to be done correctly? Think about what happens if an error is made (some delay of service, work must be redone, danger to patients or co-workers, etc.). (1–Little or no, to 5–Extremely important).
- **Difficulty:** How difficult is it to do this task correctly compared with all other tasks in the job? (1–Much easier, to 5–Much harder)

---

[20] Levine et al., "A Methodology for Developing and Validating Minimum Qualifications."

[21] Both examples are from Levine et al., "A Methodology for Developing and Validating Minimum Qualifications."

- **Criteria to be in the domain for MQs:** Majority rate Yes on both Yes/No scales, score 3 or higher on the Correct Performance, 2 or higher on Difficulty.

Examples of scales applied to KSAs for defining domains of MQs are:

- **Necessary at Entry:** Is it necessary for newly hired employees to possess this KSA on being hired or after a brief orientation/training period? (Yes/No)
- **Barely Acceptable:** Must even barely acceptable employees possess the level or amount of this KSA to do the job? (Yes/No)
- **Useful in Hiring:** To what extent is this KSA useful in choosing and hiring new employees? (1–None or very little, to 5–to an extremely great extent).
- **Unsatisfactory Employees:** How well does this KSA distinguish between the barely acceptable and the unsatisfactory employee? (1–None or very little, to 5–to an extremely great extent).
- **Criteria to Be in the Domain for MQs:** Majority rate Yes on both Yes/No scales, score 2 or higher on Useful and Unsatisfactory scales, and Useful plus Unsatisfactory must equal 5.0 or higher.

The tasks and KSAs meeting the predetermined criteria are used to form the domain of tasks and KSAs on which to base MQs. Job analysts then prepare a draft set of MQ profiles. Each profile is a statement of education, training, and/or work experience presumably needed to perform a target job at a satisfactory level. The draft MQ profiles are reviewed by a different group of SMEs. Finalized MQ profiles are then rated on two scales: Level and Clarity. The profiles meeting the criteria on these scales are then compared back to the domain of tasks and KSAs by means of two additional scales. Profiles making the cut become the new MQs for the job.[22] A joint crediting system could continue to use minimum qualifications as now but could incorporate additional profiles (beyond the now-used time and billet approach) to assess qualifications.

---

[22] Ibid, pp. 1012–1016.

Results of the study indicate that this methodology produces valid MQs. Scales applied to tasks, KSAs, and MQs profiles were also found to have acceptable validities. Tryouts of the validated MQs showed a high inter-rater reliability. In general, applying the scales resulted in a smaller task and KSA domain that included only the most important tasks and KSAs and captured the core of the job.[23] So, one course for evaluating joint experience and education for reserve active-status list officers would be to create different minimum qualification statements.

Point Method

The point method is a selection method that measures applicants by assigning points for the months or years of specified experience, education, or training. For example, if the military were to employ the point method, officers being assessed for "joint" qualification would be given a score based on points for length of JPME and joint duty assignment. The traditional point method essentially measures the time spent in education and training. It is a frequently used evaluation method in organizations.

The benefit of this method is that it is an easy and inexpensive way to assess personnel. The drawbacks, however, include a low mean validity and lack of generalizability. Meta-analyses assigned the point method a mean validity of .11 and a standard deviation of .24.[24] The variance is not surprising given that the point method scoring technique tends to give substantial weight to job experience. One would expect the validity to vary with respect to the mean level of job experience in the sample. As noted above, samples with low mean levels of job experience will show a stronger correlation between experience and performance than samples with high mean levels of job experience.

---

[23] Levine et al., "A Methodology for Developing and Validating Minimum Qualifications," pp. 1017–1019.

[24] McDaniel et al., "A Meta-Analysis of the Validity of Methods for Rating Training and Experience in Personnel Selection."

A variation of the traditional point method, the "Improved Point Method" examines an applicant's specific job behaviors/duties as indicators of job-related KSAs. In the improved point questionnaire, job incumbents identify activities that applicants could have performed that would indicate their proficiency with each job-relevant knowledge, skill, or ability.[25] This method measures experience at the *task* level rather than the *job* level. For example, analyzing a regional tasking in the context of the strategic environment for a combatant commander's AOR may be a task that demonstrates an officer's proficiency in the joint planning processes. Officers would indicate their level of experience with each activity or task and receive one point for each activity they have performed. As a result of measuring experience at the task level of specificity, the method has greater validity.

More recently, McGonigle and Curnow developed a modified improved point questionnaire that uses alternate scoring procedures to increase the validity of experience questionnaires.[26] The modified improved questionnaire measures experience performing specific behaviors that demonstrate various levels of proficiency across the performance dimensions of a job. It provides differential amounts of points based on the number of times the activity was performed. It also does not provide additional points to the applicants who exceed the point of mastery for each activity. For example, if performances in a particular war game were found to reach its peak after playing five games, the officer would receive increasing points up to the fifth game, but no extra points for additional games. The modified improved point questionnaire incorporates both of Quinones's findings that experience has greater correlation with performance when measured at the *task* level of specificity and the *amount* measurement mode. The modified improved point questionnaire also includes features to encourage

---

[25] Timothy P. McGonigle and Christina C. Curnow, "Development of a Modified Improved Point Method Experience Questionnaire," *Applied H.R.M. Research,* Vol. 7, No. 1, 2002, pp. 15–21.

[26] Ibid.

truthful responding, such as a requirement to list an individual who can verify the applicant's amount of experience with each activity. This serves to decrease faking and improve validity.

Grouping Method

The grouping method is a variation of the point method; it classifies applicants into qualification categories such as "well qualified," "qualified," and "not qualified" based on evaluation of training, education, and experience. Rather than ranking applicants over a continuous range of scores, applicants assigned to each group are given the same score and assumed to be equally suited for employment. If this method is applied to the reserve component, officers could be divided into groups of "joint qualified," "minimally joint qualified," and "not joint qualified." Currently officers are grouped into "joint qualified" and "not joint qualified."

Similar to the traditional point method, the benefit of this evaluation method is that it is easy and inexpensive to conduct using resumes or standard applications. A drawback to this approach is that studies have found relatively low inter-rater reliability coefficients for the grouping method.[27] This may be attributable to the lack of clarity in the particular rating scheme or the lack of evaluation experience on the part of the evaluators. Past research found a correlation of .82 between degree of agreement between evaluators and length of evaluation experience for the grouping method.[28]

Improved clarity in rating scheme and use of experienced evaluators should improve reliability of the grouping method.

---

[27] Ronald A. Ash and Edward L. Levine, "Job Applicant Training and Work Experience Evaluation: An Empirical Comparison of Four Methods," *Journal of Applied Psychology*, Vol. 70, No. 3, 1985, pp. 572–576.

[28] Levine and Flory, 1975 cited in Ash and Levine, 1985

Employment Tests

Employment tests are used by a variety of professions to assess candidates and measure different areas such as mental ability, achievement, and personality dimensions. Like other measures, a test is a useful tool only if it is valid and reliable.

Achievement Tests

Achievement tests include knowledge tests and work sample or performance tests. Knowledge tests typically involve specific questions to determine how much the individual knows about particular job tasks. Licensing/certifying exams are examples of knowledge tests that are used in many fields, including health care, accounting, real estate, and social work. An achievement test measuring joint competencies could be constructed for reserve officers. The test could include questions on the 33 learning objectives for Phase I JPME and 8 learning objectives for Phase II JPME.[29]

A work sample test is an achievement test that requires individuals to actually demonstrate or perform one or more tasks. A work sample for a joint officer may be to create a joint operation plan for a war in a particular region. Knowledge and work sample tests have high validities and generally show a high degree of job relatedness. Test takers tend to view these tests as fairer than other types of tests. A negative aspect of achievement tests is that they can be expensive to develop and administer.[30]

Assessment Centers

In an assessment center approach, candidates are assessed with a wide variety of instruments and procedures. This can include interviews, ability and personality measures, and a range of standardized management activities and problem-solving exercises. In-basket tests ask can-

---

[29] Dayton S. Pickett, David A. Smith, and Elizabeth B. Dial, *Joint Professional Military Education for Reserve Component Officers*, McLean, VA: Logistics Management Institute, 1998.

[30] U.S. Department of Labor, Employment and Training Administration, "Testing and Assessment: An Employer's Guide to Good Practices," March 1999, p. 42.

didates to sort through a manager's in-basket of letters, memos, directives, and reports that describe problems and scenarios.[31] Candidates are asked to examine them, prioritize them, and respond appropriately with memos, action plans, and problem-solving strategies. Leaderless group discussions are group exercises in which a group of candidates is asked to respond to various kinds of problems and scenarios without a designated group leader. Candidates may be evaluated on their teamwork, leadership, and other job-relevant skills. A role-play exercise is when candidates are asked to pretend that they already have the job and must interact with another employee (generally a trained assessor) to solve a problem.

An advantage of assessment centers is that they apply the *whole-person* approach to personnel assessment. They can be very good predictors of job performance and behavior when the tests and procedures making up the assessment are constructed and used properly.[32] The disadvantages are that they can be expensive to administer and develop. Additionally, specialized training is required for assessors. Their skills and experience are essential to the quality of the evaluations they provide. The United States Foreign Service uses an achievement test and the assessment center approach for selecting Foreign Service Officers.

Self-Rating Evaluation Methods

Some methods are based on self-ratings. A major hurdle in designing valid self-rating methods is that they are highly susceptible to faking. Applicants may answer questions untruthfully to present themselves in a more positive manner.

Task Method

The task method evaluates applicants on the basis of their experience with job-specific tasks. It is a self-evaluation in which applicants rate their experience and skill at each task. The task method can vary in

---

[31] Ibid, p. 46.

[32] Ibid.

terms of the type of self-reported data it requests and the scoring procedure that is implemented.[33] Some task method questionnaires ask applicants to assess the relative time spent performing each task. For example, "How many hours or days have you spent creating joint operation plans?" Others ask applicants to assess their performance on a task ranging from "unacceptable" to "outstanding." For example, "Please rate your performance on analyzing the role that C4I plays in joint operational planning." Still others use scales that measure the amount of supervisory assistance or additional training one needs to perform a task. For example, "Are you able to independently plan for the employment of joint forces at the operational level of war?" In a given questionnaire, all tasks may be given the same weight or some tasks may be weighted more than others. The task method is unique in that it is the only typology that is based entirely on self-ratings.

There are several benefits to employing the task method as a means of personnel selection. This method measures experience at the *task* level of specificity that has the greatest validity in measuring job performance. If tasks included in the questionnaire are reflective of types of tasks that are performed on the job, this method is a good measure of job preparedness.

A disadvantage to the task method is that it has a relatively low mean validity of .15 and a standard deviation of .27.[34] Although the mean validity of the task method is higher than that of the point method, the task method does meet the 90 percent reliability value criterion.[35] The task method is also not generally appropriate for entry-level positions, as it assumes experience conducting job tasks.

The validity of the task method can be increased under four conditions: (a) expectations of self-evaluation verification, (b) self-evaluation

---

[33] M. A. McDaniel, F. L. Schmidt, and J. E. Hunter, "A Meta-Analysis of the Validity of Methods for Rating Training and Experience in Personnel Selection," *Personnel Psychology,* Vol. 41, No. 2, 1988, p. 285.

[34] Ibid.

[35] Reliability coefficients from .90 to 1 are considered excellent. Reliability from .80 to .89 is considered good. Reliability from .70 to .79 is considered adequate. Coefficients below .70 may limit applicability. Taken from "Testing and Assessment: An Employer's Guide to Good Practices," U.S. Department of Labor, 3/99, p. 33.

instructions using social comparison terminology, (c) self-evaluation experience, and (d) anonymity of individual self-reporting.[36] Typically, the latter three conditions are absent when self-ratings of task performance are collected. While social comparison instructions (e.g., "how do you perform compared with others?") could be easily incorporated into task questionnaires, most task studies don't appear to use them. Of course, self-assessment experience is rarely under the control of the employer, and since task ratings are used for personnel selection, the applicant's self-ratings cannot be anonymous.

Studies show that use of proper methodology in developing task-based questionnaires (TBQs) can increase the validity of the task method. Steps to developing a TBQ include identifying tasks, developing a scoring system, and pilot testing.[37] The TBQ should include tasks that are critical and needed-at-entry and be limited to tasks that qualified applicants could have experience performing. To determine if an applicant is faking, the TBQ should include counterfeit items. Reflecting the higher validity associated with using the measurement mode "amount" in evaluating experience, the scoring system should award points for the number of times the task has been performed until the point of diminishing returns. TBQ developers should estimate where this occurs, based on job analysis and minimum qualifications data and assign points linearly below that. For example, points could be assigned for different levels:

- 0 points for experience that does not match needed performance
- 1 point for experience that does not match needed performance
- 2 points for experience that matches 40 percent of needed performance
- 3 points for experience that matches 60 percent of needed performance

---

[36] Mabe and West, 1982 cited in McDaniel et al., "A Meta-Analysis of the Validity of Methods for Rating Training and Experience in Personnel Selection."

[37] Patrick J. Curtin, Deborah L. Whetzel, and Kenneth E. Graham, "Identifying and Developing Predictors of Job Performance," presented at IPMAAC Conference, Baltimore, MD, 2003, p. 41.

- 4 points for experience that matches 80 percent of needed performance
- 5 points for experience that matches 100 percent of needed performance

Pilot testing is also necessary to examine readability of the TBQ and difficulty in identifying counterfeit tasks.

The greatest challenges in developing TBQs are the issues of faking, collecting accurate and reliable information, setting performance standards, and measuring education/training-based experience.[38] A three-layer protection to faking includes a signed certification of information accuracy, including penalties for falsification, verifiers (such as references that can support accuracy of information), and counterfeit items. Counterfeit items include nonsensical tasks that applicants could not perform, and multiple opportunities to "correct" responses can be included to help identify fakers.

Several steps can be taken to address the challenge of collecting accurate and reliable information. Measuring the number of times a task has been performed is the best indicator of experience, but it is often difficult to estimate. It is better to attempt to calculate the frequency and duration of a task rather than the raw number of times a task has been performed. Providing anchoring examples can also be helpful in collecting accurate information.

Estimating when performance standards are reached for tasks is complicated by several factors. First, some tasks may not have standards, including interpersonal KSAs. Second, factors besides experience, including education and training, can influence the rate of achievement. Third, estimating the rate does not account for individual differences among people, including intelligence and openness to experience.[39]

Education/training experience is hard to standardize and measure using TBQs. Factors such as difference in schools, instructors, and course materials affect the education experience. Collecting informa-

---

[38] Ibid, 45.

[39] Ibid, 42.

tion on activities performed as part of education or training may be more useful to predicting job performance than number of months or years of training.[40]

KSA-Oriented Methods

In KSA-oriented methods, such as KSA-based questionnaires (KSABQ), applicants indicate their experience in performing activities related to the job. KSABQs measure experience as indicators of KSAs and focus on quantity of experience.

A major benefit of this method is that applicants do not need to have experience with specific job tasks. KSA-based questionnaires can be used for entry level jobs and require only a minimal level of written communications skills. KSA-oriented methods have validities as high as .43.[41]

Similar to TBQs, KSABQs should be generated using a proper methodology. Steps to creating effective KSABQs include identifying qualifying KSAs, generating activities, and developing a scoring system.[42] Qualifying KSAs include those most important and those needed-at-entry. The KSABQ should be limited to those KSAs that applicants could develop proficiency in through related experience. A KSA for joint officers may include knowledge of national military capabilities and command structure. "Activities" are behavioral representations of KSAs. They should include behaviors that represent different levels of proficiency with KSAs.[43] Activities should be able to be performed as part of "feeder" jobs or through education or training. The scoring system should award points for increasing amounts of experience performing tasks. The scoring system should also reflect the asymptotic relationship between experience and performance.

---

[40] Ibid, 52.

[41] Tim McGonigle, "Theory Behind Training and Experience Measures," presented at IPMAAC Conference, Baltimore, MD, June 2003, p. 11.

[42] Tim McGonigle, "Development of Entry-Level Experience Questionnaires," presented at IPMAAC Conference, Baltimore, MD, June 2003, p. 25.

[43] Ibid, 29.

The challenges in developing KSABQs are similar to the challenges of TBQs. These include faking, collecting reliable information, setting performance standards, and measuring education/training-based experience. These challenges can be addressed in the same manner as discussed in the TBQ section.

Behavioral Consistency/Accomplishment Records

Accomplishment records are a measure of job-related previous experience and focus on "quality" of experience. In accomplishment records, applicants write major accomplishments that demonstrate their level of proficiency in several job-related areas.[44] Job-related areas are those behavioral dimensions rated by experienced supervisors as showing maximal differences between superior and minimally acceptable performers.[45] For example, relevant job-related areas for joint officers may include "mastery of principles of combined arms operations" or an "understanding of how to plan for employment of joint forces at the operational level of war." Applicants' accomplishment statements are evaluated using anchored rating scales for which the anchors are accomplishment descriptors whose values along a behavioral dimension have been reliably determined by SMEs.[46]

Accomplishment records are based on the behavioral consistency principle, which states that the best predictor of future performance is past performance in a similar circumstance.[47]

The benefits of using accomplishment records in personnel assessment include high validity and decreased susceptibility to faking. A meta-analysis of achievement records assigned a mean true validity of .45 and a standard deviation of .1. One reason that validity may be

---

[44] Tim McGonigle, "Theory Behind Training and Experience Measures," presented at IPMAAC Conference, Baltimore, MD, June 2003, p. 12.

[45] McDaniel et al., "A Meta-Analysis of the Validity of Methods for Rating Training and Experience in Personnel Selection."

[46] Ibid.

[47] Schmidt, Caplan, et al., 1979.

higher for accomplishment records than for the task method is because an applicant is less likely to fake. It is more difficult to write well-developed fabrications than it is to check a box on a task inventory. The drawbacks of accomplishment records are that they require significant written communication skills. Additionally, a good scoring guide is necessary to ensure that accomplishment records are evaluated reliably. Past studies have found accomplishment record reliabilities ranging from .75 to .85.[48]

Accomplishment records are most commonly used to select applicants for professional positions that require experience. They have been used most frequently to select attorneys. Other professions that use achievement records are administrative law judges and teachers.

---

[48] Hough, 1984; Hough et al., 1983; Sadowski & Hess, 1994, cited in McGonigle and Curnow.

Bibliography

Ash, Ronald A., James C. Johnson, Edward L. Levine, and Michael A. McDaniel, "Job Applicant Training and Work Experience Evaluation in Personnel Selection," *Research in Personnel and Human Resources Management,* Vol. 7, 1989, p. 199.

Ash, Ronald A., and Edward L. Levine, "Job Applicant Training and Work Experience Evaluation: An Empirical Comparison of Four Methods," *Journal of Applied Psychology,* Vol. 70, No. 3, 1985, pp. 572–576.

Booz Allen Hamilton, *Independent Study of Joint Officer Management and Joint Professional Military Education,* McLean, VA, 2003.

Butler, Shelly, "Other Variables Affecting T&E Measures," presented at IPMAAC Conference, Baltimore, MD, June 2003.

Capstone Concept for Joint Operations, Version 2.0, Washington, D.C.: Director for Operational Plans and Joint Force Development, Joint Staff J-7, Joint Experimentation Transformation and Concepts Division, Pentagon, August 2005.

Chairman of the Joint Chiefs of Staff Instruction, Officer Professional Military Education Policy (OPMEP), CJCSI 1800.01C, December 22, 2005.

CJCS Guide 3501, "The Joint Training System: A Primer for Senior Leaders," October 10, 2003.

CJCS Vision for Joint Officer Development, November 2005.

Congressional Research Service, *Department of Defense Reorganization Act of 1986: Proposals for Reforming the Joint Officer Personnel Management Program,* Washington D.C.: July 18, 2000.

Curtin, Patrick J., Deborah L. Whetzel, and Kenneth E. Graham, "Identifying and Developing Predictors of Job Performance," presented at IPMAAC Conference, Baltimore, MD, 2003, p. 41.

Davenport, LTC Judith A., "An Analysis of Reserve Component Joint Officer Management Including Five Major Issues and Suggested Recommendations," March 13, 2004, unpublished.

Department of Defense Directive, Military Training, Number 1322.18, September 3, 2004.

Department of Defense Directive 1300.19, as of November 2003.

Department of Defense Instruction, Reserve Component (RC) Joint Officer Management Program, Number 1215.20, September 12, 2002.

Director, RC JPME, JFSC, Information Paper on Joint Professional Military Education (JPME) for the Reserve Components (RC), dated August 25, 2004, unpublished.

Draft Air Force Personnel Information Systems Concept of Operations (as of July 14, 2004), http://www.icodap.org/HRRD/CONOPS-HRRD.htm (as of March 2006).

Driscoll, John B., "Developing Joint Education for the Total Force," *Joint Forces Quarterly,* Spring 2000, pp. 87–91.

Hanson, Marshall, "IMA Reservist," The Officer: Naval Services—Service Sections, http://www.findarticles.com/p/articles/mi_m0IBY/is_3_80/ai_ 115405159/print . . . (as of January 31, 2006).

Joint Training Policy for the Armed Forces of the United States, Joint Staff, CJCSI 3500.01B, December 31, 1999.

Kirby, Sheila Nataraj, Al Crego, Harry J. Thie, Margaret C. Harrell, Kimberly Curry, and Michael S. Tseng, *Who Is "Joint"?: New Evidence from the 2005 Joint Officer Management Census Survey,* Santa Monica, Calif: RAND Corporation, TR-349-OSD, 2006.

Levine, Edward L., Doris M. Maye, Ronald A. Ulm, and Thomas R. Gordon, "A Methodology for Developing and Validating Minimum Qualifications (MQs)," *Personnel Psychology,* Vol. 50, No. 4, 1997, pp. 1009–1024.

McDaniel, Michael A., Frank L. Schmidt, and John E. Hunter, "Job Experience Correlates of Job Performance," *Journal of Applied Psychology,* Vol. 73, No. 2, 1988, p. 330.

McDaniel, Michael A., F. L. Schmidt, and J. E. Hunter, "A Meta-Analysis of the Validity of Methods for Rating Training and Experience in Personnel Selection," *Personnel Psychology,* Vol. 41, No. 2, 1988, pp. 285–314.

McGonigle, Tim, "Theory Behind Training and Experience Measures," presented at IPMAAC Conference, Baltimore, MD, June 2003.

McGonigle, Timothy P., and Christina C. Curnow, "Development of a Modified Improved Point Method Experience Questionnaire," *Applied H.R.M. Research,* Vol. 7, No. 1, 2002, pp. 12–21.

National Defense Authorization Act for FY 2005.

National Guard Bureau, *National Guard 2005 Posture Statement,* March 2004.

Paauwe, J., and P. Boselie, *Challenging (Strategic) Human Resource Management Theory: Integration of Resource-Based Approaches and New Institutionalism,* Rotterdam, The Netherlands: Erasmus Institute of Management (ERS-2002-40-ORG), April 2002.

Pickett, Dayton S., David A. Smith, and Elizabeth B. Dial, *Joint Professional Military Education for Reserve Component Officers,* McLean, VA: Logistics Management Institute, 1998.

Quinones, Miguel A., J. Kevin Ford, and Mark S. Teachout, "The Relationship Between Work Experience and Job Performance: A Conceptual and Meta-Analytic Review," *Personnel Psychology,* Vol. 48, No. 4, 1995, p. 891.

Thie, Harry J., Margaret C. Harrell, and Robert M. Emmerichs, *Interagency and International Assignments and Officer Career Management,* Santa Monica, Calif: RAND Corporation, MR-1116-OSD, 1999.

Thie, Harry J., Margaret C. Harrell, et al., *Framing a Strategic Approach for Joint Officer Management,* Santa Monica, Calif: RAND Corporation, MG-306-OSD, 2005.

Thie, Harry J., Roland J. Yardley, Peter Schirmer, Rudolph H. Ehrenberg, and Penelope Speed, *Factors to Consider in Blending Active and Reserve Manpower Within Military Units,* Santa Monica, Calif.: RAND Corporation, MG-527-OSD, forthcoming.

U.S. Air Force Armstrong Laboratory, Summary of Data Bases, December 1992, http://www.icodap.org/040510/SUMMARY1992.htm#n.%20UOR (as of March 2006).

U.S. Air Force Personnel Center, Military Modernization, http://www.afpc. randolph.af.mil/dlearn/milmod/default.htm (as of March 2006).

U.S. Department of Defense, *Strategic Plan for Joint Officer Management and Joint Officer Development* April 3, 2006.

U.S. Department of Labor. See http://www.dol.gov/odep/pubs/fact/analysis. htm.

U.S. Department of Labor Employment and Training Administration, "Testing and Assessment: an Employer's Guide to Good Practices," March 1999, pp. 3–10.

U.S. General Accountability Office, *Performance Measurement and Evaluation: Definitions and Relationships*, GAO-05-739SP, May 2005.

U.S. Government Accountability Office, *Military Personnel: Joint Officer Development Has Improved, but a Strategic Approach Is Needed*, GAO-03-238, December 2002.

U.S. Government Accountability Office, *Human Capital: A Guide for Assessing Strategic Training and Development Efforts in the Federal Government*, GAO-03-893G, July 2003.

U.S. Government Accountability Office, *Military Training: Actions Needed to Enhance DOD's Program to Transform Joint Training*, GAO-05-548, June 2005.

U.S. Government Accountability Office, *Force Structure: Assessments of Navy Reserve Manpower Requirements Need to Consider the Most Cost-effective Mix of Active and Reserve Manpower to Meet Mission Needs*, GAO-06-125, October 2005.